660.28449
dc 2

# Fermentation Biotechnology

# The Biotechnology Series

This series is designed to give undergraduates, graduates and practising scientists access to the many related disciplines in this fast developing area. It provides understanding both of the basic principles and of the industrial applications of biotechnology. By covering individual subjects in separate volumes a thorough and straightforward introduction to each field is provided for people of differing backgrounds.

## Titles in the Series

**Biotechnology: The Biological Principles:** M.D. Trevan, S. Boffey, K.H. Goulding and P. Stanbury

**Fermentation Kinetics and Modelling:** C.G. Sinclair and B. Kristiansen (Ed. J.D. Bu'lock)

**Enzyme Technology:** P. Gacesa and J. Hubble

**Animal Cell Technology:** M. Butler

**Fermentation Biotechnology:** O.P. Ward

**Genetic Transformation in Plants:** R. Walden

## Upcoming Titles

Biotechnology of Biomass Conversions    Plant Biotechnology in Agriculture

Biosensors                                   Applied Gene Technology

Biotechnology Safety                      Bioreactors

Plant Cell and Tissue Culture

## Overall Series Editor

Professor J.F. Kennedy    *Birmingham University, England*

## Series Editors

Professor J.A. Bryant    *Exeter University, England*

Dr R.N. Greenshields    *Biotechnology Centre, Wales*

Dr C.H. Self    *Hammersmith Hospital, London, England*

## The Institute of Biology I**O**B

*This series has been editorially approved by the **Institute of Biology** in London. The Institute is the professional body representing biologists in the UK. It sets standards, promotes education and training, conducts examinations, organizes local and national meetings, and publishes the journals **Biologist** and **Journal of Biological Education**.*

*For details about Institute membership write to: Institute of Biology, 20 Queensberry Place, London SW7 2DZ.*

# Fermentation Biotechnology

## Principles, Processes and Products

*Owen P. Ward*

*Open University Press*

*Milton Keynes*

*to Alice, Conor, Evelyn and David*

Open University Press
Open University Educational Enterprises Limited
12 Cofferidge Close
Stony Stratford
Milton Keynes MK11 1BY

First Published 1989

**British Library Cataloguing in Publication Data**
Ward, Owen P.
  Fermentation biotechnology : principles,
  processes, products.
  1. Industrial fermentation. Microbiological
  aspects
  I. Title   II. Series
  660.2'8449

  ISBN 0-335-15172-8
  ISBN 0-335-15171-X Pbk

Typeset by Vision Typesetting, Manchester
Printed in Great Britain by Oxford University Press

# Contents

# Preface

In so many ways, the fermentation process represents an essential link or bridge. It links the ancient arts of making cheese, wine and oriental fermented foods using natural microbial flora with a modern food fermentation industry using pure cultures and sophisticated process control equipment. It links ancient practices of producing algae for food consumption in alkaline ponds to large scale, recently developed, single cell protein production processes. Successful development of the fermentation process has enabled Alexander Fleming's observations of the antagonistic effects of *Penicillium notatum* on *Staphyloccus aureus* to ultimately evolve into a major antibiotic industry. Many other important historic examples exist. More recently, fermentation has emerged as a vital, pivotal technology, or integrating force, in modern biotechnology. Development of fermentation technology depends on the skills of molecular and cellular biologists and process engineers. It links genetic engineering to commercially viable large scale industrial processes. The spectacular development of hybridoma technology and commercialization of the first plant cell culture process, have resulted in the fermenter becoming a production tool, not only for manufacture of microbial products, but for mammalian and plant cell products as well. The requirement for finer and more automated fermentation process control and the development of biological detection devices or biosensors is linking biology to electronic and computer technology. Expansion in commercial activity over the coming decades, arising from research in molecular and cellular biology, assures fermentation biotechnology of a bright future. In the longer term, fermentation of renewable raw materials may replace depleting non-renewable fossil-fuels as a source of bulk chemicals. This book, focusing on principles, processes and products, attempts to embrace all of these perspectives.

A special acknowledgement is due to my colleagues, Dr C.W. Robinson,

Department of Chemical Engineering and Dr B.R. Glick, Department of Biology, University of Waterloo and to my former colleague, Dr M. Clynes, School of Biological Sciences, N.I.H.E., Dublin, each of whom reviewed sections of the book and provided expert advice and criticism. Thanks are also due to my colleagues, Val Butler, Kathy Clarke, Kitty Hain and Shelley Stobo who assisted in preparation of the manuscript.

*Owen P. Ward*

# Acknowledgements

I wish to thank the authors, publishers and manufacturing companies listed below for granting permission to reproduce original or copyright material.

*Authors*
Abson, J.W. & Todhunter, K.H. (Figure 12.1); Bauer, K. (Figure 10.7); Furuya, A. *et al.* (Figure 9.6); Greenshields, R.N. (Figure 7.8); McGregor, W.C. (Figure 5.13); Miller, T.L. & Churchill, B.W. (Table 4.2); Ng, T.K. *et al.* (Figure 8.1); Pirt S.J. (Figure 6.7); Porubscan, R.S. & Sellars, R.L. (Figure 6.8); Queener, S.W. & Swartz, R.W. (Figure 10.4).


*Publishers and manufacturing companies*
Academic Press Inc., Orlando, Florida: Figure 6.8 from Porubscan, R.S. & Sellars, R.L. (1979). Lactic Starter Culture Concentrates. *Microbial Technology*, Vol. 1, 59–61; Figure 12.1 from Abson, J.W. & Todhunter, K.H. (1967). Effluent Disposal. *Biochemical and Biological Engineering Science*, Vol. 1, 310–343; Figure 7.8 from Greenshields, R.N. (1978). Acetic Acid: Vinegar. *Economic Microbiology*, Vol. 2, 121–186; Figure 10.4 from Queener, S.W. & Swartz, R.W. (1979). Penicillins: Biosynthetic and Semisynthetic. *Economic Microbiology*, Vol. 3, 35–123.
Agricultural Chemical Society of Japan: Table 10.2 from Fujita, Y. and Hara, Y. (1985). *Agric. Biol. Chem.* 49 (7), 2071–2075.
American Chemical Society: Figure 5.14 from Pandrey, R.C. *et al.* (1985). Process Developments in the Isolation of Largomycin F-II, a Chromoprotein Antitumour Antibiotic. *Purification of Fermentation Products*, ACS Symposium Series 271, 133–153.
American Society for Microbiology: Figure 9.6 from Furuya A. *et al.* (1968).

Production of nucleic acid related substances by fermentative processes. XIX. Accumulation of 5'inosinic acid by a mutant of *Brevibacterium ammoniagenes*. *Applied Microbiology*, 16, 981–987; Table 4.2 from Miller, T.L. & Churchill, B.W. (1986). Substrates for Large-Scale Fermentations. *Manual of Industrial Microbiology and Biotechnology*, 122–136.

American Association for the Advancement of Science: Figure 6.3 from Litchfield, J.H. (1983). Single-cell proteins. *Science*, 219, 740–746; Figure 8.1 from Ng, T.K. et al. (1983). Production of feedstock chemicals. *Science*, 219, 733–740.

Ametek/Process Equipment, California: Figure 5.6.

Bio/Technology: Figure 3.12 from Wilson, T. (1984). Bioreactor, synthesizer, biosensor markets to increase by 16 percent annually. *Bio/Technology*, 2, 869–873; Figure 5.1 from Dwyer, J.L. (1984). Scaling up biproduct separation with high performance liquid chromatography. *Bio/Technology*, 2, 957–964; Figures 5.2 and 5.12 from Fish, N.M. & Lilly, M.D. (1984). The interactions between fermentation and protein recovery. *Bio/Technology*, 2, 623–627; Figure 10.14 from Van Brunt (1986b). Immobilized mammalian cells: the gentle way to productivity. *Bio/Technology*, 4, 505–510; Figure 10.15 from Posillico, E.G. (1986). Microencapsulation technology for large-scale antibody production. *Bio/Technology*, 4, 114–117; Figure 10.16 from Klausner, A. (1986). 'Single chain' antibodies become reality. *Bio/Technology*, 4, 1041–1043; Table 10.1 from Ratafia, M. (1987). Mammalian cell culture: worldwide activities and markets. *Bio/Technology*, 5, 692–694.

Blackwell Scientific Publications, Oxford: Figures 2.1 and 2.5 from Deacon, J.W. (1984). *Introduction to Modern Mycology*.

British Mycological Society: Figure 8.7 from Bu'Lock, J.D. et al. (1974). Regulation of secondary biosynthesis in *Gibberiella fujikuori*. *Transactions of the British Mycological Society*, 62, 377–389.

Butterworth Scientific Ltd., England: Table 9.4 from Sinden K.W. (1987). The production of lipids by fermentation within the EEC. *Enzyme and Microbial Technology*, 9, 124–125.

Cambridge University Press: Figure 6.6 from Anderson, C. & Solomons, G.L. (1984). Primary metabolism and biomass production from *Fusarium*. *The Applied Mycology of Fusarium*, 231–250; Figure 7.1 from Hough, J.S. (1985). *The Biotechnology of Malting and Brewing*.

Churchill Livingstone, Edinburgh: Figures 10.8, 10.9, 10.10, 10.11, 10.12 and 10.13 from Van Hemert, P. (1974). Vaccine Production as a Unit Process. *Progress in Industrial Microbiology*, Vol. 13, 151–271.

CRC Press Inc., Boca Raton, Florida: Table 6.1 from Solomons, G.L. (1983). Single Cell Protein. *Critical Reviews of Biotechnology*, 1 (1), 21–58.

Edward Arnold, London: Figure 3.7 from Kristiansen, B. and Chamberlain, H.E. (1983). Fermenter Design. *The Filamentous Fungi*, Vol. IV, 1–19.

Gulf Publishing Co., Texas: Figure 6.2 from Laine, B.M. (1974). What proteins cost from oil. *Hydrocarbon Processing*, 53, (11), 139–142.

Intercept Ltd., England: Figures 3.10, 12.2 and 12.3 from Wheatley, A.D. (1984). Biotechnology of Effluent Treatment. *Biotechnology and Genetic Engineering Reviews*, Vol. 1. 261–310.

John Wiley & Sons Inc., New York: Figure 6.5 from Bernstein, S. *et al.* (1977). The Commercial Fermentation of Cheese Whey for the Production of Protein and/or Alcohol. *Single Cell Protein from Renewable and Nonrenewable Resources, Biotechnology and Bioengineering Symposium*, 7, 1–9.

Leonard Hill Books, London: Figure 5.5 from Purchas, D.B. (1971). *Industrial Filtration of Liquids.*

McGraw Hill Book Company, New York: Figure 5.3 from Bailey, J.E. & Ollis, D.F. (1986). *Biochemical Engineering Fundamentals, 2nd Edition.*

New York Academy of Science: Figure 5.13 from McGregor, W.C. (1983). Large-Scale Isolation and Purification of Recombinant Proteins from Recombinant *E. coli. Annals of the New York Academy of Science*, 413, 231–236.

Penwalt Corporation, Sharples-Stokes Division, Pennsylvania: Figure 5.4.

Pergamon Press Ltd., Oxford: Figure 5.9 in Tutunjian, R.S. (1985). Ultrafiltration Processes. *Comprehensive Biotechnology*, Vol. 2, 411–437; Figure 7.6 in Irving, D.M. & Hill, A.R. (1985). Cheese Technology. *Comprehensive Biotechnology*, Vol. 3, 523–565; Figure 9.3 in Nakayama, K. (1985). Lysine. *Comprehensive Biotechnology*, Vol. 3, 607–620; Figures 10.18 and 10.19 from Flickinger, M. (1985). Anticancer Agents. *Comprehensive Biotechnology*, Vol. 3, 231–273.

Springer-Verlag, New York: Figure 2.14 from Demain, A.L. (1971). Overproduction of Microbial Metabolites and Enzymes Due to Alteration of Regulation. *Advances in Biochemical Engineering*, Vol. 1, 113–142.

Van Nostrand Reinhold (UK) Ltd: Figure 2.17 from Priest, F.G. (1984). *Extracellular Enzymes.* Figure 3.8 and 3.9 from Smith, J.E. (1985). *Biotechnology Principles.*

# Chapter 1

# *Introduction*

**Fermentation as an ancient art**

In broad terms, fermentation involves the use of micro-organisms to carry out enzyme-catalysed transformations of organic matter. Fermentation has been performed as an art for many centuries. Wine-making is thought to have been practised at least since 10 000 BC. Historians believe that the Egyptians produced beer in 5000–6000 BC. Barley, which had been germinated in earthenware vessels, was crushed, mixed as a dough and baked, and finally soaked in water and allowed to produce beer. About 4000 BC the Egyptians used brewer's yeast for carbon dioxide production in the leavening of bread. Algae of the genus *Spirulina* were harvested for food consumption from alkaline ponds by the ancient Aztecs in Mexico. The origins of a variety of indigenous fermented foods and sauces in the Orient and elsewhere, now known to be based on fermentation, enzyme production and enzymatic hydrolysis using surface culture methods, clearly also go back many thousands of years. Records of the transformation of milk into products such as cheese date back to 5000 BC, when it was observed that milk carried in calf stomachs (which contain milk-clotting enzymes) tended to curdle. Vinegar has probably been known for as long as wine-making has been practised, although the earliest records referring to vinegar are in the Old and New Testaments. Earliest references to distilled potable spirits date back to 1000BC in China.

   Production of modified foods and beverages involving fermentation have been operated for approximately 10 000 years before the existence of micro-organisms was recognized. There is also evidence that gradually process improvements were made in these traditional technologies. Microscopic examination of sediments of excavated beer urns dating from 3400 BC and 1440 BC almost certainly indicate yeast cells and it has been claimed that the yeast in the more recent sediment was of improved purity.

On the pharmaceutical side mouldy cheese, meat and bread had been employed in folk medicine for thousands of years to heal wounds and treat infections. In retrospect, the beneficial effect of such treatments was undoubtedly due to their antibiotic activity.


## The modern era of industrial fermentation technology

MICROBIAL TECHNOLOGY

Antonie van Leeuwenhoek, a pioneering Dutch microscopist, first observed yeast cells when he examined drops of fermented beer with a primitive microscope in 1680. His discovery of the yeast cell was, however, soon forgotten and fermentation continued, as it had been before, to be almost exclusively studied by chemists, who did not consider fermentation to involve living material. Then in 1836-7 three researchers, Cagniard-Latour, Schwann and Kutzing, independently stated their opinions that yeast was a living 'thing'. These reports were ridiculed by a number of eminent chemists including Berzelius, Wohler and von Liebig. In 1847, Blondeau, a physics professor, made a study of fermentations involving lactic acid, butyric acid, acetic acid and urea and seems to have been first to state that different fermentations were carried out by different 'fungi'. Despite these observations it was not until 1856-7 that Louis Pasteur, having carried out detailed investigations on beer and wine fermentations, finally concluded that living yeast cells ferment sugar into ethanol and carbon dioxide, when they are obliged to live in the absence of air. Pasteur also studied a large number of other fermentations. He noted that an organism (probably a *Penicillium* sp.) selectively fermented *d*-ammonium tartrate in a racemic mixture of *d,l*-tartrate. He observed how cylindrical organisms produced butyric acid only under anaerobic conditions and also investigated the production of acetic acid by fermentation. In the 1870s Pasteur, together with a number of other researchers had observed the antagonistic effects of one micro-organism on another and predicted potential therapeutic applications. Another important milestone in the development of industrial and medical microbiology dates back to 1881 when Robert Koch, a medical doctor, pioneered the development of pure culture techniques and other classical bacteriological methods which are used to this day.

Two vital elements set the stage for the modern era of industrial fermentation technology—the traditional art of using yeasts and moulds for food and beverage modification and the pioneering microbiological studies of scientists like Pasteur and Koch. The technology for surface or semi-solid fermentation processes had developed through production of oriental foods. The ability of the moulds to hydrolyse starch and protein, for example in soy sauce production, prepared the way forward for industrial enzyme production. Anaerobic fermentations, particularly production of beverage alcohol, had been developed and characterized to some extent. The cultivation of algae and yeast paved the way for industrial processes for production of microbial food protein (single-cell protein, SCP) and microbial inoculants. The traditional production of vinegar and Pasteur's more recent studies on production of acids prepared the path for the development of fermentations to produce organic acids. The development of pure culture methodology by Koch provided others with the techniques to study the industrial applications of individual microbial strains and set the scene for the use of pure cultures, sterilized media and aseptic conditions in industrial fermentations.

## ANIMAL AND PLANT CELL CULTURE

Although techniques for growing mammalian cells *in vitro* have been practised for a century, developments in this area have in general taken place more recently than those of microbial technology. Roux performed the first 'explantation' experiments, using living tissue from chick embryo in 1885. Around 1910, Harrison and Carrel developed many of the classical mammalian cell-culture techniques. They refined the salt and amino acid composition of growth media, used supplemental plasma and clotted lymph as growth-promoting additives for the first time and developed bioreactors for the propagation of viruses and other products. In 1950 Morgan and colleagues developed the first chemically-defined medium capable of supporting growth of chick embryo for 4–5 weeks. Between 1955 and 1960, Eagle and co-workers defined many of the nutritional requirements of cultured mammalian cells. The specific vitamin and amino acid requirements determined by this group for reproducible growth of L-cells have served as the basis for the majority of current strain-specific cell-culture formulations. These developments enabled viral vaccines to be produced using cell-culture techniques from the 1950s and prepared the way for major expansion in the industrial use of cell culture for production of monoclonal antibodies and other products which occurred in the mid-1980s.

The first successful culture of plant cells was achieved by Gautheret in 1934. During the 1950s and 1960s, plant cell-culture techniques had developed to a stage where applications in horticulture and agriculture for strain development were widespread. The large-scale cultivation of tobacco and a variety of vegetable cells was examined in the late 1950s and early 1960s. Interest developed in the use of plant cells in culture to produce high-value secondary-metabolite plant products. In 1985, Mitsui Petrochemical Company, Japan, started production of shikonin by *Lithospermum erythrorhizon*, the first case of commercial production of a plant secondary metabolite by cell culture.

## FOOD FERMENTATIONS

While modern industrial food fermentations have developed from traditional fermented food processes, the food fermentations themselves have also continued to develop technologically. Processes such as bread-making, cheese manufacture, brewing and distilling have been developed to meet modern commercial requirements of large-scale production, high and consistent quality, competitive costs and product variety. Highly-standardized preparations of baker's yeast and dairy starter cultures and the use of industrial enzymes have improved efficiency, automation and quality control of modern baking and dairy fermentation processes.

Bread dough fermentations have been accelerated by use of a higher proportion of yeast at higher temperatures. Use of microbial amylases liberates fermentable sugars from the starch grains, thus providing sugar for the yeast to ferment to $CO_2$ bubbles which raise the dough and give the bread its characteristic texture. Microbial proteases are often used to partially hydrolyze wheat gluten proteins in doughs, thereby improving dough-handling properties, increasing loaf volume and improving loaf shape.

The use of cheese starter cultures is one of several factors which have contributed to the development of a wide variety of high-quality cheese products. Cheese-making no longer relies on spontaneous milk infection. Processes have been streamlined by use of starter cultures and other process aids in combination with modern milk-handling equipment. Serious problems of bacteriophage infection of starter cultures were encountered. Measures taken to overcome these problems included use of starter culture mixtures, rotation of phage-unrelated starter strains and use of phage inhibitors. Since the beginning of the 1980s bacteriophage-insensitive strains have been successfully developed.

Since Pasteur's time many developments have taken place in the alcoholic beverage industries. Pasteurization allowed trade in beer to develop from a localized business to national and international scale. The brewing, distilling and wine-making industries made a rapid transition from small craft-like units to large complexes where the aim is to maintain a consistent product even though raw materials, plant and scale of operation are changing. In order to achieve a reproducible product, fermentations are standardized by controlling parameters such as inoculation or pitching rate, yeast-cell viability, yeast storage conditions, dissolved oxygen concentration at pitching, soluble nitrogen and wort-fermentable sugar concentrations, and fermentation temperature. Fermentation innovations such as use of continuous fermentation processes, brewing at high wort gravity, use of unmalted cereals and microbial enzymes and production of low carbohydrate beer have been used with varying levels of success. Conventional genetic techniques have been applied to improve desirable yeast properties and eliminate undesirable characteristics.

## BAKER'S YEAST AND SINGLE-CELL PROTEIN

Commercial bakers obtained their yeast supplies from local breweries in the seventeenth century. Due to its bitter taste and variable fermentation activities, brewer's yeast was gradually replaced by distiller's yeast, which in turn was replaced by baker's yeast. Pressed baker's yeast was first produced around 1781 using the so-called Dutch process and later in 1846 using the Vienna process which produced low yeast yields of 5% and 14%, respectively, based on raw material, and concurrent yields of spirits of around 30%. Substantial progress was made in yeast biomass industrial fermentations between 1879 and 1919 leading to a baker's yeast industry independent of alcoholic beverage production. Aeration of the grain mash was introduced by Marquardt in 1879, which increased the yield and reduced accompanying spirit production to 20%. Further refinements involving incremental sugar feeding, the first example of fed-batch fermentation processes, raised the production efficiency close to theoretical maximum with no concomitant formation of alcohol.

During World War I in Germany, baker's yeast, grown on molasses, was produced as a protein supplement for human consumption—the first time modern fermentation processes were used for SCP production. Later in Germany during World War II, *Candida utilis*, cultivated on sulphite waste liquor from pulp and paper manufacture and sugar derived from acid hydrolysis of wood, was used as a protein source for humans and animals. This process has since been used in the USA, Switzerland, Taiwan and the USSR. From 1968 onwards several companies in Europe, the USA and Japan built SCP plants. Some of these plants have since closed down whether because of regulatory problems or high production costs, while other companies are continuing their development programmes. In 1981, ICI in England, scaled up its SCP production, using the aerobic bacterium *Methylophilus methylotrophus* to produce 3000 t/month of the product Pruteen for animal feed. The first novel microbial protein produced by fermentation to receive government approval for human use, Mycoprotein, is produced by the fungus *Fusarium graminearum* by Rank Hovis McDougall in England. The filamentous morphology of Mycoprotein confers on it a natural texture similar to meat, enabling the product to command a realistic market price and gain consumer appeal.

## CHEMICALS AND FOOD ADDITIVES

*Organic acids*
Commercial production of lactic acid by fermentation began in 1881 and today 50% of industrial lactic acid is produced by fermentation using *Lactobacillus delbruckii*.

Citric acid was first produced by extraction from lemon juice and later synthesized from glycerol and other chemicals. In 1923 citrate was produced by industrial fermentation. Ten years later, annual world production exceeded 10 000 t with greater than 80% produced by fermentation. Annual current market estimates are more than 350 000 t, exclusively produced by fermentation. Initially, surface culture methods were used with *Aspergillus niger* as producing organism. Submerged *A. niger* processes were introduced after World War II and around 1977 a more efficient submerged process involving a *Candida* species was commercialized.

Other organic acids, which are economically manufactured by fermentation, include gluconic and itaconic acid. Food-grade acetic acid (vinegar) is produced exclusively by oxidation of ethanol using *Acetobacter aceti* but all industrial-grade acetic acid is now manufactured solely by chemical methods.

*Alcohols and ketones*

During World War I supplies of imported vegetable oils, used for the production of glycerol, were cut off from Germany and a fermentation process was quickly developed to produce glycerol for use in explosives' manufacture. The process was based on the discovery by Neuberg that glycerol is produced by yeast, at the expense of ethanol, in the presence of sodium bisulphite. The British satisfied a war requirement for acetone by development of the anaerobic acetone–butanol fermentation involving *Clostridium acetobutylicum*. This was the first large-scale fermentation requiring pure-culture methods to prevent contamination. After World War I the demand for acetone declined but the need for *n*-butanol increased and it continued to be produced by fermentation until the 1950s when the price of petrochemicals dropped below that of starch and sugar-based substrates. The acetone–butanol fermentation was operated commercially until very recently by National Chemical Products, South Africa, where petroleum was scarce due to international embargoes.

Chemically-produced industrial alcohol was extensively used in the USA from the late 1800s. The industrial alcohol fermentation industry began in the USA following the repeal of prohibition and in 1941 accounted for 77% of the industrial alcohol market. During the 1950s and 1960s, ethylene was cheap and efficient hydration processes were developed for its conversion to ethanol making synthetic ethanol production more cost competitive than fermentation methods. The 1973 oil embargo and the resultant oil price increases generated world-wide interest in alternative renewable fuel systems. Brazil embarked on a national alcohol programme aimed at developing the fermentation capacity to produce 3 billion gallons of ethanol annually from cane-sugar by 1987. The USA federal government encouraged the development of alcohol production from corn by fermentation in the Gasohol Program. The excise tax on gasohol (10% alcohol in gasoline) was reduced and a considerable number of small- and large-scale fermentation plants were established to produce alcohol using both batch- and continuous-fermentation processes. The Program target was to achieve an annual output of 1.8 billion gallons by the mid-1980s. Ethanol is also produced by fermentation of whey by *Kluyveromyces fragilis* although the volume of ethanol produced in this way is tiny compared to the volume produced by *S. cerevisiae* from corn or cane-sugar. Whey fermentations must be carried out close to the cheese factory as transport costs of whey, which only contains about 5% w/v lactose, are prohibitive.

Clearly, only a small number of commodity chemicals are currently produced by fermentation. Practically all commodity chemicals are manufactured from petroleum and natural gas. The fluctuating cost and uncertain supply of petroleum and concerns about ultimate depletion of non-renewable resources have intensified interest in use of non-petroleum feedstocks for chemical and energy production.

## Amino acids

During the last thirty years, production of amino acids by aerobic fermentation processes has rapidly expanded. Monosodium glutamate and lysine are produced in largest amounts with annual world production levels of 370 000 t and 40 000 t, respectively. Fermentation processes are the result of elegant research work on the biochemical mechanisms which regulate microbial amino acid biosynthesis. Amino acid over-production has been achieved by a combination of mutations and fermentation process control. The techniques used aim to stimulate cell uptake of new materials, to prevent or hinder side reactions, to induce and activate biosynthetic enzymes, to reduce and inhibit enzyme activity which would degrade the product and finally to facilitate excretion of the amino acid. The classical research approach used to develop the glutamic acid fermentation has provided significant impetus to the development of fermentation processes for production of other primary metabolites. As a result almost all the amino acids are now produced commercially by fermentation. Similar techniques have been applied to production of the flavour enhancers inosine and guanosine monophosphates.

## Biopolymers

A number of industrial fermentations have been developed in recent years for production of microbial biopolymers. Most polymers are synthetic and sensitive to the cost of petroleum feedstocks so that biopolymers will be especially important if oil prices go up.

The most important biopolymer in terms of fermentation production volume is xanthan gum, which is used as a gelling agent or suspension stabilizer. This gum is produced by aerobic fermentation of the bacterium *Xanthomonas campestris*. Other microbial gums of commercial interest are produced by *Azotobacter vinelandii* (alginate), *Aureobasidium pullulans* (pullulan), *Sclerotium* species (scleroglucan) and *Pseudomonas elodea* (gellan). The high molecular weight and high viscosity of extracellular polysaccharide gums present problems for fermenter mixing and mass and heat transfer, which must be taken into account in development of an optimal fermenter design for these processes.

A very promising polymer currently being developed is polyhydroxybutyrate, a thermoplastic polyester which can accumulate intracellularly in some species of *Alcaligenes eutrophus* to a level of up to 70% of biomass weight. This product would enable biodegradable plastic to be used in place of non-biodegradable petroleum-based plastics.

## Vitamins

Most vitamins are currently produced by chemical methods. While fermentation methods have been described for a number of group B vitamins, such as thiamine, biotin, folic acid, pantothenic acid, pyridoxine, vitamin $B_{12}$ and riboflavin, only the last two are produced by biological methods. Chemical synthesis of vitamin $B_{12}$ is extremely complicated and it is produced commercially solely by industrial fermentation using *Pseudomonas* strains. With riboflavin production fermentation procedures compete effectively with chemical synthesis or semi-synthesis methods. Thirty percent of the riboflavin world market is supplied by fermentation. In 1935, *Eremothecium ashbyii* was first used to produce the vitamin and yields are reported to have been gradually increased to 5.3 g/l. Because of strain instability another *Ascomycetes* species, *Ashbya gossypii*, later became the preferred strain. A recent patent (1984) describes a recombinant strain of *Bacillus subtilis* capable of producing 4.5 g riboflavin in a 24-h fermentation, which is much shorter than the *Ascomycetes* processes which take about five days.

*Microbial insecticides*

Chemical insecticides, which have been eminently successful in agricultural and public health applications, have been widely used this century. However, criticisms that they may kill non-target organisms or that target organisms may become resistant have led to the development and commercialization of microbial insecticides and ongoing research is continuing in this area. Production of *Bacillus thuringiensis* far exceeds any other commercially-produced microbial insecticide. Fermentation conditions are designed to achieve optimal yield and bioactivity of the crystal endotoxin which is produced concomitantly with sporulation. Different strains of *B. thuringiensis* and indeed other bacterial and fungal insect pathogens exhibit different toxicities. *Bacillus thuringiensis* toxin genes have been cloned into *Escherichia coli* and *B. subtilis* and this provides potential for improvement in rates of production and for alteration of the target spectrum.

*Gibberellins*

Gibberellins, which are diterpenoid compounds containing four cyclic carbon rings, have application in the acceleration of barley germination in malt production. The first gibberellin was discovered in 1938. Sixty different compounds are now known to exist. Gibberellic acid was originally produced by surface culture in an extended fermentation to achieve yields of 40–60 mg/l product. Commercial fermentations producing significantly higher yields are now carried out by submerged culture.

MICROBIAL ENZYMES

Commercial exploitation of isolated microbial enzymes dates back to Jokichi Takamine, a native of Japan who emigrated to the USA, and who in 1894 patented a method for preparation of diastatic enzymes from moulds, and which was marketed under the name Takadiastase. His method, involving growth of the mould on the surface of a solid substrate, such as wheat or bran, undoubtedly reflected his insight into similar processes for preparation of oriental fermented foods. The development of bacterial industrial enzyme fermentations was pioneered by Boidin and Effront, from France and Germany, respectively, whose 1917 patent described the use of *Bacillus subtilis* and *Bacillus mesentericus* to produce amylases and diastases, again using surface cultivation techniques. Surface culture methods for fungal enzyme production were used in the USA well into the 1950s and continue to be used, especially in Japan, to the present day. Following the submerged fermentation experience gained through the development of the penicillin production process, at the Northern Regional Research Laboratories (NRRL) in the USA, submerged fungal fermentations for enzyme production were developed in the USA and in Europe.

The introduction of industrial microbial enzymes to the marketplace lay a foundation for the development of enzyme application technology, particularly in food processing areas such as brewing, production of fruit juices and manufacture of hydrolysed starch syrups. In the 1960s, the market for industrial enzymes expanded dramatically following the inclusion of alkaline proteases in washing powder preparations. An important milestone in the 1970s involved the production and use of microbial glucose isomerase, for manufacture of high-fructose corn syrup (HFCS). This had a major impact on the caloric sweetener market, dominated up till then by sucrose. In the USA, the beverage industry is the largest industrial user of sugar. By 1984 all major soft drinks' companies there had approved total substitution of sucrose by HFCS. Another significant technical development of the 1970s was the commercial production of high-temperature stable α-amylase from *Bacillus licheniformis*. The ability of this enzyme to liquefy starch at temperatures up to 110°C has not only streamlined starch-hydrolysis processes but has also created an impetus to isolate or construct other enzymes with high stability properties.

## HEALTH CARE PRODUCTS

### Antibiotics

For fifty years, following the suggestion of Pasteur that the antagonistic effects of one micro-organism or another might have therapeutic potential, various microbial preparations were tested as medicines without success. Finally, in 1928 Alexander Fleming observed that *Penicillium notatum*, when present as a contaminant in culture dishes of *Staphylococcus aureus*, killed the bacterium. He showed that the active ingredient, named penicillin, could inhibit many bacteria. Following isolation of penicillin in a stable active form by Florey and Chain around 1940, its remarkable antibacterial activity was demonstrated and a commercial fermentation was developed with the assistance of the NRRL in the USA and the involvement of a number of US pharmaceutical companies.

Thus the successful development of the penicillin fermentation marked the start of the antibiotic industry. Soon afterwards a number of new antibiotics, including streptomycin, were isolated from *Streptomyces* species and cephalosporins were shown to be produced by *Cephalosporium acremonium*. While the penicillins, cephalosporins and streptomycins were the most important early antibiotic discoveries, hundreds of new antibiotics have been isolated annually from the 1940s right up to the present day.

The development of the penicillin fermentation process also led to some key general advances in industrial fermentations. It marked the beginning of efficient submerged fungal fermentation processes. The golden age of classical microbial genetics, involving techniques of mutation and selection, which started around 1945, enabled yields of penicillin, for example, to be increased from a few milligrams per litre to over $20\,g/l$ of culture. Studies were also initiated on the nature of the complex metabolic processes involved in the production of secondary metabolites, the group of small molecules which includes antibiotics, and which has no obvious role in growth and maintenance of the cell.

Modified (semi-synthetic) penicillins with an altered spectrum of activity or altered sensitivity to acid or enzyme inactivation were developed initially by chemical alteration of naturally-produced penicillin. Later, some of the semi-syntheses were achieved using biological conversion steps.

*Steroid transformations*
The use of micro-organisms to carry out highly specific and selective enzymatic transformation reactions of pharmaceuticals represented a major breakthrough in the development of steroid hormonal medicines. In the 1940s it was established that cortisone, a steroid secreted by the adrenal gland, could relieve pain of patients with rheumatoid arthritis. A chemical synthesis procedure, requiring 37 reaction steps was developed leading to production of cortisone at a cost of $200/g. In 1952 scientists at the Upjohn Company discovered a strain of *Rhizopus arrhizus* which could hydroxylate progesterone to produce 11-$\alpha$-hydroxyprogesterone. This enabled the synthesis procedure for cortisone to be shortened from 37 to 11 steps and reduced the cost to $16/g. Process improvements have continually been made over the years, progressively lowering the cost of production so that by 1980 cortisone cost $0.46/g.

Similar microbial biotransformation techniques have been applied to synthesis of other steroid drugs, particularly aldosterone, prednisone and prednisolone. The techniques are also important for synthesis of other pharmaceutical and non-medical chemicals.

*Vaccines*
There are references to infectious diseases in ancient Hindu writings and there is archaeological evidence of tuberculosis lesions in mummified remains taken from early Egyptian tombs.

The ability to confer resistance to disease by vaccination was first described by Jenner, in 1798, who observed that individuals inoculated with cowpox or 'vaccinia' were immune to subsequent challenge with smallpox.

The use of vaccines and immunology had its beginnings about 1877, when Pasteur turned his attention to the causes and prevention of infective diseases of man and animals. In 1890 Behring and Kitasato succeeded in immunizing animals with inactivated diphtheria and tetanus toxins. Following isolation of *Vibrio cholerae* by Koch, the first cholera vaccine was administered in 1885 by Ferran Y. Clua, a Spanish physician, by subcutaneous injection of live broth cultures into more than 30 000 Spaniards. Vaccination programmes for typhus were conducted in the armies during World War I and similar programmes using tetanus were introduced in World War II. Mass vaccinations using diphtheria vaccine started in the late 1930s. Widespread use of pertussis and poliomyelitis vaccines followed reports on the success of field trials with these vaccines in 1955. Calmette and Guérin around 1936 observed that serial cultivation of virulent or infective tubercle bacilli on artificial media for long periods attenuated the culture to the point where it could no longer cause disease. Three years later the first BCG vaccine (Bacille of Calmette and Guérin) was introduced for immunization against tuberculosis. Use of attenuated vaccines received a major setback ten years later in the terrible Lübeck disaster, in which 72 of 240 BCG-vaccinated infants died as a result of treatment with a batch of vaccine containing virulent tubercle bacilli. Bacterial vaccines used today consist of live attenuated strains of the disease carrying organisms or closely-related strains, killed pathogenic strains or cell components containing antigens which are effective in inducing immunity to target infectious diseases. While vaccine production is an important application of fermentation, its scale of manufacture is relatively small. Fermentation technology involves batch or continuous processes and conditions are designed to optimize cellular production of the immunogenic material.

Viral vaccine technology was primitive for many years because of the need for animals as a source of virus. Use of chick embryos was then established as a source of viruses and later, around 1950, use of tissue culture cells was introduced. Thus with the development of cell-culture techniques, viral vaccine production entered the industrial fermentation arena. Mammalian cells are cultured in an appropriate medium and then at a certain stage of growth, the culture is inoculated with the virus, which multiplies in the mammalian cell host. Cell culture enables large quantities of virus to be prepared with much less contamination by extraneous material from host tissue, contaminating bacteria or viruses than occurs in the earlier methods.

## Impact of hybridoma and recombinant DNA technologies

Important recent developments in hybridoma technology and genetic engineering have extended the scope and potential of industrial fermentation technology.

### MONOCLONAL ANTIBODIES

In 1975 Kohler and Milstein fused a mouse myeloma (skin cancer cell) with an antibody-producing white blood cell to make a hybrid (hybridoma) cell which combined the distinctive cell-dividing and antibody-producing capabilities of the parent cells. They developed the basic technology for production of monoclonal antibodies (MCAs), preparations of individual specific antibodies obtained from cells derived from a single ancestor or clone. MCAs have high binding specificity for individual receptors on a molecule or surface and there are many existing and potential applications, which take advantage of this unique characteristic particularly in clinical analysis and disease therapy. The market for MCAs has grown rapidly and is expected to reach $1 billion by 1990. Monoclonal antibody production has been carried out *in vivo* by injection of the hybridoma clone into the abdominal cavity fluid of the animal (usually mouse) or *in vitro* in a variety of cell-culture systems. Greater overall productivity is achieved in cell-culture systems and a variety of cell reactor systems have been developed and are used commercially.

## RECOMBINANT DNA TECHNOLOGY

Genetic engineering involves the formation of new combinations of heritable material by insertion of foreign genes, produced outside the cell, into a host organism in which they do not naturally occur. The first gene was cloned in 1973 and since then techniques have been developing which can currently or will soon enable 'foreign' proteins to be produced in commercial quantities by a variety of recombinant strains of prokaryotes and eukaryotes, including animal and plant cells. The pharmaceutical industry has invested heavily in this R & D activity and the first therapeutic product to achieve regulatory approval was human insulin, produced by a recombinant strain of *E. coli*. A variety of other proteins have been produced in recombinant strains, including other human hormones and growth factors, anti-tumour and anti-viral compounds, viral antigens and enzymes. The first genetically-engineered vaccine to be marketed was a veterinary vaccine for pseudorabies, a herpes virus that infects swine, and which was introduced in 1986.

Initial pioneering genetic engineering research and the first products commercialized used *E. coli* as host since its genetic systems were more fully understood. Workhorses of the fermentation industry, *Bacillus*, *Aspergillus* and *Saccharomyces* spp. may be better hosts in the longer term once their genetic systems are more fully understood. Proteins are generally not secreted by *E. coli* whereas *Bacillus* spp. produce high yields of extracellular proteins. Since prokaryotes do not glycosylate proteins, *Saccharomyces* and *Aspergillus* species may become target hosts for production of animal or human glycoproteins. *Aspergillus niger* is known to be capable of secreting up to 20 g/l protein from a single extracellular enzyme gene. Some recombinant pharmaceuticals are being produced using *S. cerevisiae*. However, protein synthesis and secretion systems are significantly more complex in these eukaryotic micro-organisms than in prokaryotes and have to be more fully investigated to enable these organisms to be used in fermentation, particularly for production of recombinant non-pharmaceutical products, for example, enzymes.

Proteins and peptides, the translation products of structural genes, are the first obvious targets for production using recombinant DNA technology. Longer term, it may be possible to manipulate primary and secondary metabolism to improve metabolite production by genetic engineering of key enzymes in the metabolic sequence.

More specific aspects of this topic are discussed in the individual chapters which follow.

# Chapter 2

# Biology of industrial micro-organisms

## Introduction

Commercially important products of industrial fermentations fall into four major categories—microbial cells, large molecules such as enzymes and polysaccharides, primary products and secondary metabolites which are not needed for growth. Cells used to produce these products exhibit a variety of physiological and biochemical properties. Commercial production of fermentation products has mainly involved use of a range of species of bacteria, yeasts and fungi. Recent advances in the development of techniques for cultivation of animal and plant cells has enabled these more complex cells to be used also in fermentation processes. In this chapter the properties of the major species involved are discussed and general aspects of microbial growth and metabolism relevant to industrial cell-culture processes are considered.

## Industrial micro-organisms

Two major classes of cells, prokaryotes and eukaryotes, exist in nature and both types are used in industrial fermentation processes. Eukaryotic cells have a distinct nucleus surrounded by a membrane. Nuclear DNA is associated with protein and exists as definite structures called chromosomes. The cells also contain other structures or organelles having specific physiological or biochemical functions. In contrast, prokaryotes lack a well-defined nucleus so that the genetic material in the form of double-stranded DNA is not separated from other cell constituents by its own membrane. These cells also lack other specialized organelles present in eukaryotes, such as mitochondria and the enzymes associated with these

**Table 2.1**  Selected characteristics which differentiate between prokaryotes and eukaryotes

| Characteristic | Prokaryotes | Eukaryotes |
|---|---|---|
| *Organelles* | | |
| Nucleus | − | + |
| Mitochondria | − | + |
| Endoplasmic reticulum | − | + |
| *Genome* | | |
| No. of DNA molecules | $1^1$ | $>1$ |
| DNA in organelles | − | + |
| DNA as chromosome structures | − | + |

1. Bacteria may contain small DNA fragments (plasmids) in addition to the single major genome.

organelles in eukaryotes are found in the protoplasm and plasma membrane of prokaryotes. The cellular properties of prokaryotes and eukaryotes are compared in Table 2.1. Bacterial cells, which are prokaryotic, and fungal, yeast, animal and plant cells, which are eukaryotic, are used in fermentation processes.

Micro-organisms are also distinguished based on their oxygen requirements. Strict aerobes, such as *Streptomyces* and most filamentous fungi, can metabolize and grow only in the presence of atmospheric oxygen. Strict anaerobes, such as *Clostridia*, can only grow in the absence of oxygen. Facultative organisms, including industrial yeasts, can grow aerobically or anaerobically.

Bacteria involved in fermentation processes are mainly chemo-organotrophs, that is, they obtain their energy and carbon by oxidation of organic compounds. Some properties of the most important genera involved and some example species are listed in Table 2.2. Bacteria are distinguished as Gram-positive or Gram-negative based on their response to Gram-staining, which binds to peptidoglycan. The cell envelope of Gram-positive bacteria consists largely of peptidoglycan, 15–80 nm thick, covering a single cytoplasmic membrane. In Gram-negative cells, this peptidoglycan layer is only 2–3 nm thick. Gram-negative cells have two membranes, an outer membrane and cytoplasmic membrane, separated by a periplasmic space. These different structures are illustrated in Fig. 2.1. Some prokaryotes, including the important antibiotic-producing *Streptomyces*, have filaments or mycelia.

Bacterial reproduction occurs by asexual cell division as shown diagrammatically for *Eschericia coli* in Fig. 2.2. The single circular DNA chromosome is replicated and a septum or dividing membrane develops in the middle of the cell, separating the new and old chromosome. The septum develops into a wall and the cells separate. Spores are produced by spore-forming bacteria usually in response to unfavourable environmental conditions. Spores are more resistant to heat and toxic chemicals than vegetative cells. When bacterial spores are transferred to appropriate media they germinate to reform vegetative cells.

The fungi are also chemo-organotrophs which have branched filamentous rigid

**Table 2.2** Properties of chemo-organotrophic bacterial genera of importance in industrial fermentations

| Genus | Gram reaction | Endo-spores | Shape | Aerobic[1] | Example species (products) |
|---|---|---|---|---|---|
| *Enterobacteria* | − | − | Rod | ± | *E. coli* (recombinant proteins) |
| *Pseudomonas* | − | − | Rod | + | *P. ovalis* (gluconic acid), *P. fluorescens* (2-oxogluconate) |
| *Xanthomonas* | − | − | Rod | + | *X. campestris* (xanthan gum) |
| *Acetobacter* | − | − | Rod | + | *A. aceti* (acetic acid) |
| *Alcaligenes* | − | − | Rod/coccus | + | *A. faecalis* (curdlan gum) |
| *Bacillus* | + | + | Rod | + | *B. subtilis*, *B. licheniformis*, *B. amyloliquefaciens* etc. (extracellular enzymes), *B. coagulans* (glucose isomerase), *B. licheniformis* (bacitracin), *B. brevis*, (gramicidin S), *B. thuringiensis* (insecticide) |
| *Clostridium* | + | + | Rod | − | *C. acetobutylicum* (acetone, butanol), *C. thermocellum* (acetate) |
| *Leuconostoc* | + | − | Spherical | ± | *L. mesenteroides* (dextran) |
| *Lactobacillus* | + | − | Rod | − or ± | *L. delbruckii* (lactic acid) |
| *Corynebacterium* | + | − | Irregular | + | *C. glutamicum* (amino acids), *C. simplex* (steroid conversion) |

**Table 2.2** continued

| Genus | Properties | | | | Example species (*products*) |
|---|---|---|---|---|---|
| | Gram reaction | Endo-spores | Shape | Aerobic[1] | |
| *Mycobacterium* | + | − | Branching filaments No mycelia | + | *Mycobacterium* spp. (steroid transformation) |
| *Nocardia* | + | − | Fragmenting mycelia | + | *N. mediterranei* (rifamycins), *Nocardia* spp. (steroid transformation) |
| *Streptomyces* | + | − | Intact mycelia | + | *S. erythreus* (erythromycin), *S. aureofaciens*, *S. rimosus* (tetracyclines), *S. fradiae* (glucose isomerase), *S. roseochromogens* (steroid hydroxylations) |

1. Aerobic: + = aerobic; − = anaerobic; ± = both.

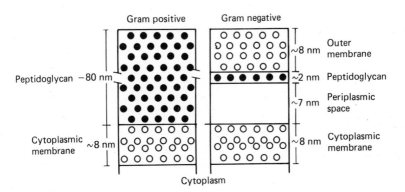

**Fig. 2.1** Schematic comparison of the cell envelope composition of Gram-positive and Gram-negative bacteria.

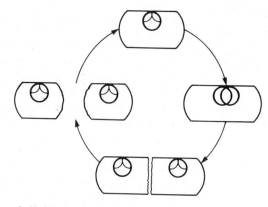

**Fig. 2.2**   Asexual division in *Escherichia coli*.

hyphae usually of diameter 2–18 μm. Higher fungi typically have cross-walls or septa at intervals along the hyphae, but these are characteristically absent from the lower fungi. The most important fungi involved in industrial fermentations are from two of the main classification groups, the aseptate Zygomycotina, which includes the *Mucor* and *Rhizopus* genera, and the septate Deuteromycotina or Fungi Imperfecti, which include the genera *Trichoderma*, *Aspergillus*, *Penicillium*, *Aureobasidium* and *Fusarium* (Fig. 2.3).

The Zygomycotina produce haploid vegetative mycelia. Asexual spores are formed in a sporangium. They also reproduce sexually. Deuteromycotina produce haploid vegetative mycelia. Asexual spores are born on conidia which vary according to species. Asexual spores are usually induced by certain environmental conditions. For example, spore-bearing conidiophores are initiated in *Aspergillus niger* when nitrogen becomes limiting.

The major fungal cell-wall constituents in the Zygomycotina are chitosan and chitin while glucan and chitin predominate in Deuteromycotina. Vegetative growth of filamentous fungi involves hyphal extension and occurs at the hyphal tip. Branching also occurs with the development of new extension zones. The length that hyphae grow and the amount of branching depends on the growth environment. Shear forces in fermenters can cause hyphal fragmentation and may result in production of more highly-branched shorter mycelia.

In fungi, the hyphal tip is the main area of protoplasm activity (Fig. 2.4). The cell wall, overlying the plasmalemma, is only 50 nm thick in the apical region. Behind the tip, the cell wall may thicken to about 125 nm and the protoplasmic compartments gradually become more vacuolated as they become more distant from the apex.

Yeasts are micro-fungi which generally exist in the form of single cells and which reproduce by budding (Fig. 2.5). Some yeasts only form individual cells and sometimes short chains while others have a range of cell forms including various filament types. A feature of a growing population of yeast cells is the presence of buds which are produced when the cell divides. The daughter cell starts as a small bud, increases until it is similar in size to the mother cell and then separates. In yeast, the cell envelope includes the cytoplasmic membrane, consisting of lipids, proteins and mannans, a periplasmic space and the cell wall, containing some protein and a large amount of glucan and mannan.

*Saccharomyces cerevisiae* is the most widely used yeast in industrial fermentations, through its applications in alcohol production and baking. *Saccharomyces cerevisiae* cannot degrade lactose and another yeast, *Kluyveromyces lactis*, which contains the necessary lactose-transporting and degrading enzymes, is used to produce alcohol or biomass from whey. Other important industrial yeasts include *Candida utilis* and *Endomycopsis fibuliger*.

Mammalian cells are generally more complex than fungal or yeast cells. Because they normally exist in controlled isotonic environments, they do not possess tough outer cell walls. The external surface of the plasma membrane contains a coat or glycocalyx consisting of short oligosaccharide chains linked to intrinsic membrane lipids and proteins and to some absorbed proteins. In tissues other than reproductive tissues, cells are diploid and cell division takes of the order of 24 h. Depending on cell type, animal cells can grow in cultures as monolayers attached to surfaces (anchorage-dependent) or as suspended cells (anchorage-independent) in slowly-rotated, gently-stirred media. Mammalian cell nutritional requirements are very complex and cells are also very sensitive to fluctuations in parameters such as temperature, pH, dissolved oxygen and $CO_2$.

Cell lines prepared directly from animal tissues generally do not survive, although this varies. They also usually exhibit contact inhibition, that is, they stop growing when they touch. However, after repeated passaging, some primary cell lines become transformed, that is, they multiply faster and grow to higher densities than primary cells, can be cultured indefinitely and lose the property of contact inhibition. Tumour cell lines are already transformed.

Hybridoma cells are constructed by fusing a myeloma cell with an antibody-producing cell, such that the hybridized cell has the characteristics of transformed cells and can synthesize a single antibody or monoclonal antibody. These cells are used both in cell culture or implanted in the peritoneum of animals to produce monoclonal antibodies.

Many plant cell types can grow on a solid medium or in suspension using techniques similar to those used for animal cell culture. In addition to the complex media requirements which include plant extracts and auxins, carbohydrate must be added to media, as photosynthesis by cells in culture is not as efficient as in whole plants. Plant tissue grows on solid media as callus, which develops as a mixture of parenchymatous cells. On repeated subculture, the cells tend to grow faster, have less complex media requirements, may change their chromosome number and be less able to differentiate. Suspension culture is the most widely used method of plant cell cultivation. The major advantage of suspension culture is that a homogeneous environment can be established throughout the reactor.

Viruses are submicroscopic agents which infect plants, animals and bacteria. They consist of genomes made up of DNA or RNA enclosed within a protein coat, containing identical subunits, which are sometimes covered by an outer membrane of lipoprotein. Viruses have no cytoplasm or cytoplasmic membrane

| Classification group | Mycelium | Spores | Genus | Drawing | Colony appearance |
|---|---|---|---|---|---|
| Zygomycotina (Order: Mucorales) | Aseptate thin cottony or felted covers of mycelium | Asexual spores formed in sporangium | Mucor | | M. miehei and M. pusillus produce grey colonies |
| | | | Rhizopus | | Grows with extreme rapidity filling petri dish with cotton-like mass of mycelium |
| Deuteromycotina (Order: Moniales) | Septate filamentous fungi forming loose mycelium on solid or liquid substrate. Trichoderma, Aspergillus and Penicillium have light coloured mycelium. Aureobasidium has dark mycelium. Fusarium has light or dark mycelium | Asexual spores, sexual reproduction rare or unknown | Trichoderma | | Most species have colonies that spread rapidly forming a somewhat thin layer of mycelium with green conidial patches |
| | | | Aspergillus | | Rapidly spreading colonies with mycelium, white at first, frequently developing bright yellow areas and producing dark brown or black conidial heads. A. oryzae colonies are yellowish to brownish-green |

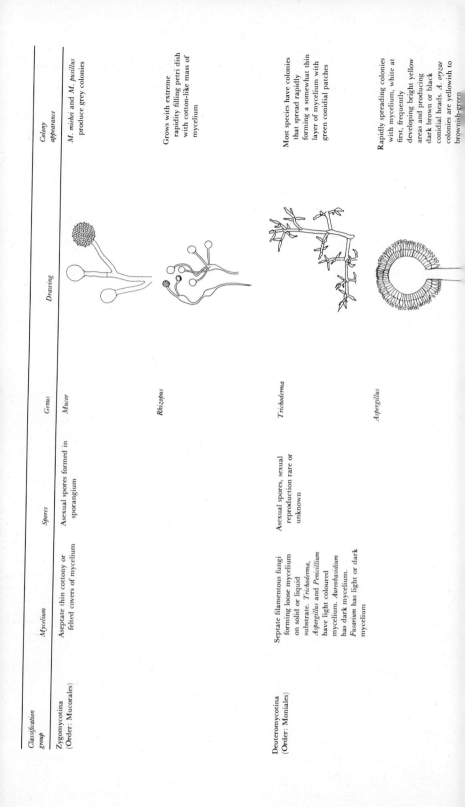

*Penicillium*

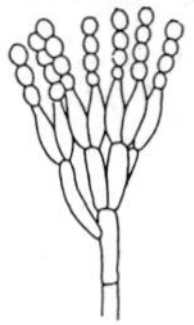

*P. chrysogenum*: Broadly spreading colonies, blue-green to bright green with broad white margin during growing period. Usually becomes greyish or purplish brown in age with overgrowth of white or rosy hyphae. **Reverse yellow**

*Aureobasidium*

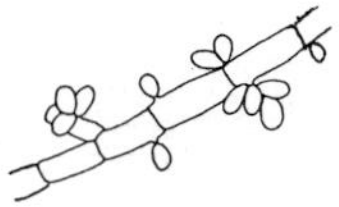

*A. pullulans*: Light to pink and brown hyphae becoming darker with age

*Fusarium*

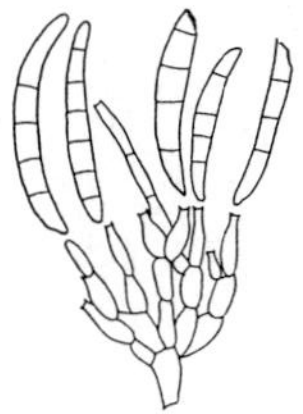

*F. graminearum*: Grey-pink to red colonies becoming brown-red with white or grey-yellow sectors

**Fig. 2.3** Characteristics of fungi of importance in industrial fermentations.

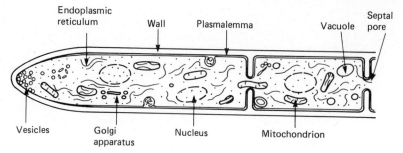

**Fig. 2.4** Diagrammatic representation of a fungal hypha (reproduced with permission from Deacon, 1984).

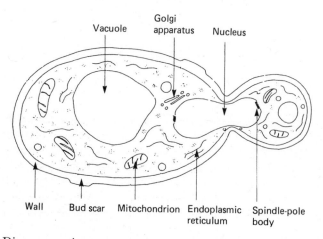

**Fig. 2.5** Diagrammatic representation of a budding yeast (reproduced with permission from Deacon, 1984).

and they replicate inside specific host cells by using the biosynthetic systems of the host under the direction of the viral genome. Viral vaccines are produced by infection of cultured host cells. Bacterial viruses (bacteriophages) which infect fermentation process microbial strains have caused serious problems particularly in cheese starter cultures.

### Cell growth

Microbial growth is an integral part of nearly all fermentation processes. We have already seen that bacteria reproduce by binary fission to produce two daughter cells of equal size. A doubling of the number of yeast cells normally results from yeast budding mechanisms. Fungi grow by hyphal extension and the morphology of the fungal mycelium can vary depending on the environment. Fungi grow on

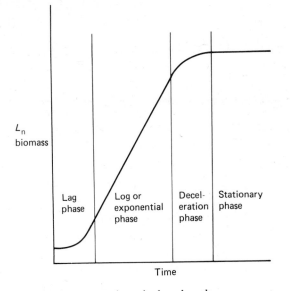

**Fig. 2.6** Growth of micro-organisms in batch culture.

the surface of solid media producing a mycelial mat, the nature of which will reflect media composition. In submerged culture, fungal cells may grow attached to suspended nutrient particles, as diffuse filamentous mycelia or as dense pellets. The morphological form of growth invariably influences rate of growth and product formation. Because of limitations in nutrient and product diffusion inside fungal pellets, growth and metabolism occur predominantly at the periphery.

When an organism is inoculated into a fixed volume of medium, the culture passes through a number of growth phases as indicated in Fig. 2.6. After inoculation there is a lag phase, where the organism takes time to adapt to the environmental conditions. Following a short period of time during which the growth rate of cells gradually increases, the cells grow at a constant maximum rate. This growth phase is called the exponential or log phase. As growth continues, nutrients are consumed and products are excreted by the organism. The growth rate decreases and finally growth ceases often due to depletion of an essential nutrient or build up of a toxic product. This phase is described as the stationary phase.

Growth measured by increase in cell mass may be described by

$$\frac{\mathrm{d}x}{\mathrm{d}t} = \mu x - \alpha x$$

(1)

where $x$ = the cell concentration ($\mathrm{mg\,cm^{-3}}$)
$t$ = the incubation time (h)
$\mu$ = the specific growth rate ($\mathrm{h^{-1}}$)
$\alpha$ = the specific rate of lysis or endogenous metabolism ($\mathrm{h^{-1}}$)

When nutrient and other cell conditions are favourable $\mu \gg \alpha$ and equation (1) becomes

$$\frac{dx}{dt} = \mu x \tag{2}$$

Integration of equation (2) gives

$$x_t = x_0 e^{\mu t} \tag{3}$$

where $x$ = cell concentration at time 0
$x_t$ = cell concentration after a time interval of t(h)
On taking the natural logarithms of equation (2) we obtain

$$\ln x_t = \ln x_0 + \mu t \tag{4}$$

If $t = t_d$, the doubling time, then $x_t = 2x$

$$\ln 2x_0 = \ln x_0 + \mu t_d \tag{5}$$

The time taken for the biomass to double, $t_d$, is given by

$$t_d = \frac{\ln 2}{\mu} \tag{6}$$

## FACTORS AFFECTING THE GROWTH RATE

Growth rate varies with microbial cell type and also in response to physical and chemical environmental conditions. In general, the cell doubling time increases with increasing organism complexity. Average doubling times for animal and plant cells are substantially longer than for yeasts and moulds which in turn are longer than for bacteria.

As with chemical and enzymatic reactions, cell growth varies as a function of temperature. Most micro-organisms will grow over a range of 25–30°C although the actual temperature at which a particular organism grows depends on its

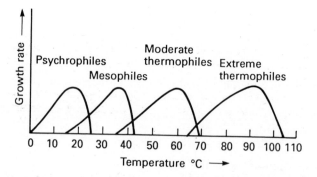

**Fig. 2.7**   Effect of temperature on specific growth rate of psychrophiles, mesophiles and thermophilic micro-organisms.

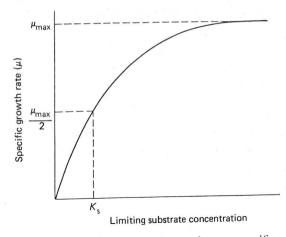

**Fig. 2.8** Effect of limiting substrate concentration on specific growth rate.

psychrophilic, mesophilic, moderate thermophilic or extreme thermophilic nature as indicated in Fig. 2.7. Animal and plant cells have much narrower temperature ranges for growth. Within the growth range, growth rate increases with increasing temperature until a maximum growth rate is reached above which growth rate falls off rapidly due to increased rate of microbial death. The effect of temperature on growth rate may be described by the Arrhenius equation

$$\mu = Ae^{-E_a/RT} \tag{7}$$

where $A$ = the Arrhenius constant
$E_a$ = the activation energy (kcal mole$^{-1}$)
$R$ = the universal gas constant
$T$ = the absolute temperature

pH influences microbial growth rate in a manner similar to its influence on enzyme activity. Most micro-organisms grow over a pH range of 3 to 4 pH units.

Growth rate of microbial cells is influenced by the water activity or relative humidity which is defined as

$$A_w = P_s/P_w \tag{8}$$

where $P_s$ is the vapour pressure of water in a solution and $P_w$ is the vapour pressure of pure water. Bacteria require an $A_w$ of $\geqslant 0.95$ whereas moulds can grow at a lower water activity of $\geqslant 0.7$.

As with any chemical reaction, growth rate will depend on the concentration of chemical nutrients and can be described by the Monod equation (Fig. 2.8)

$$\mu = \mu_{max}\frac{s}{K_s + s} \tag{9}$$

where $\mu_{max}$ = the maximum specific growth rate
$s$ = the residual substrate concentration of the limiting substrate
$K_s$ = the saturation constant and is equivalent to the substrate concentration at $\mu = 0.5\ \mu_{max}$

Typical $K_s$ values for various carbon sources are in the range $1-10$ mg dm$^{-3}$. Cells will grow at rates close to $\mu_{max}$ if the limiting carbon source is greater than $10K_s$ or $10-100$ mg dm$^{-3}$. High carbohydrate concentrations may inhibit growth due to osmotic effects which, as was seen earlier with water activity, have a more severe effect on bacteria than fungi.

It has been observed that the total amount of cell mass formed during cell growth is often proportional to the amount of substrate utilized

$$Y_{x/s} = \frac{\text{Biomass produced (g)}}{\text{Substrate consumed (g)}}$$
(10)

where $Y_{x/s}$ = the biomass yield co-efficient.


## Metabolism

PRIMARY METABOLISM

Microbial growth depends on the cell's ability to use the nutrients in its surroundings to synthesize the macromolecular components of cellular structures and also the many low-molecular-weight compounds required for cellular activity. Intermediary metabolism is concerned with the reactions which transform carbon and nitrogen compounds entering the cell either into new cell material or into products which are excreted. The synthesis of these compounds requires energy and most of the cells involved in industrial fermentations are heterotrophs which obtain this energy from breakdown of organic compounds. In aerobic or respiratory processes, organisms are able to completely oxidize some of the substrates to $CO_2$ and $H_2O$, resulting in the provision of maximum energy for conversion of the remaining substrates into new cell mass. In anaerobic or fermentative metabolism, cells are less efficient in converting organic substrates into cellular material and usually excrete partially degraded intermediates.

Energy-producing or catabolic pathways generate ATP and reduced co-enzymes needed for various biosynthetic reactions and chemical intermediates used as starting points for biosynthesis.

Sugars are broken down by one of three pathways, the Embden–Meyerhof–Parnas (EMP) pathway, the hexose monophosphate (HMP) pathway and the Entner–Doudoroff (ED) pathway. The EMP route, which converts glucose to two molecules of pyruvate, via triose phosphate intermediates, occurs most widely in animal, plant, fungal, yeast and bacterial cells. As it does not produce precursors for biosynthesis of aromatic amino acids, DNA and RNA, micro-organisms which solely use this pathway for glucose utilization must be supplied with growth factors. The ED pathway has a relatively restricted occurrence. A few bacteria, including *Pseudomonas* species, which do not meta-bolize glucose via the EMP pathway, utilize this pathway. One cycle of the HMP pathway results in the net conversion of glucose to $CO_2$ with production of 12 molecules of $NADPH + H^+$. Micro-organisms can use this pathway to produce pyruvate. The importance of the HMP pathway lies especially in provision of precursors of aromatic amino acids and vitamins and the supply of NADPH and $H^+$ needed for many biosynthetic reactions.

A significant end-product of all three pathways above is pyruvic acid, which is channelled into the tricarboxylic acid (TCA) cycle in aerobic metabolism and is likewise the precursor of the acid, alcohol and other end-products of anaerobic metabolism. A large number of micro-organisms meet their energy needs via the TCA cycle. Acetyl CoA (2C) formed from pyruvate condenses with oxaloacetate (4C) to form citrate (6C) which is converted back to oxaloacetate over the course of the cycle with production of two molecules of $CO_2$ (see Fig. 2.9a). Pyruvate is thus oxidized to $CO_2$ with formation of reduced co-enzymes which can be re-oxidized by passing their electrons along the electron-transport chain to electron acceptors such as oxygen. During this respiration process, some of the energy produced is used to form ATP needed for biosynthesis.

TCA cycle intermediates also serve as precursors for biosynthesis of many amino acids, some organic acids and other fermentation products. When cycle intermediates are diverted into biosynthesis in this way, it is necessary to restore the cycle intermediates so as to maintain the level of oxaloacetate required to condense with incoming acetyl groups from pyruvate. A number of anaplerotic (replenishing) reactions exist which serve this function. The main ones involve carboxylation of pyruvate or phosphoenol pyruvate to form oxaloacetate or malate. Micro-organisms grown on carbon sources such as fatty acids, acetate or their precursors, use the TCA cycle for production of both energy and biosynthetic precursors. Acetyl CoA is formed which condenses with oxaloacetate to enter the cycle. Oxaloacetate and malate, by being converted to phosphoenol pyruvate and pyruvate can be used as precursors of hexoses and pentoses using reversible reactions of the EMP and HMP pathways. Again in the absence of anaplerotic reactions TCA cycle intermediates would be depleted. In this case the introduction of two enzymatic reactions leads to another cyclic mechanism, known as the glyoxylate cycle, which replenishes $C_4$ acids using incoming acetyl CoA. These key anaplerotic reactions are summarized in Fig. 2.9. The glyoxylate cycle may also utilize acetyl CoA produced from hexose via pyruvate to replenish TCA cycle intermediates.

Hexoses are converted to single-cell protein (SCP) by yeasts and fungi generally by use of a combination of the EMP and HMP pathways followed by the TCA cycle and respiration. Production of SCP from alkanes involves alkane oxidation to acetate units and its assimilation by the glyoxylate and TCA cycles, with enzymes of the EMP and HMP pathway used for biosynthesis of hexoses, pentoses and other cell constituents. In production of SCP from methanol by *Methylophilus* species, the substrate is converted to formaldehyde which is in turn oxidized by the ribulose monophosphate pathway (see Fig. 6.4 on p. 100).

In the citric acid fermentation process involving *Aspergillus niger*, hexoses are converted, via the EMP pathway, to pyruvate and acetyl CoA which condenses with oxaloacetate to form citrate in the first step of the TCA cycle. Further citrate metabolism is prevented by inhibition of a TCA enzyme downstream from citrate and oxaloacetate needed for the condensation reaction is replenished by carboxylation of pyruvate. The EMP pathway and pyruvate carboxylation also

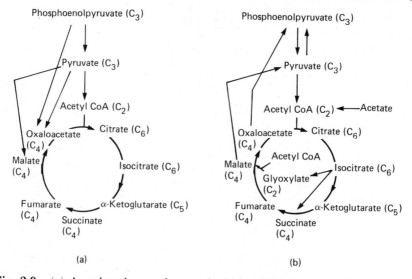

(a)                                                          (b)

**Fig. 2.9** (a) Anaplerotic reactions replenishing TCA intermediates when an organism is grown on carbohydrates. (b) Anaplerotic and other key reactions for synthesis of hexose and pentose precursors when an organism is cultured on substrates which generate acetate as a major carbon source.

feature in the process for production of itaconic acid from hexose by *A. terreus*. Production of glutamic acid and lysine by *Corynebacterium* and *Brevibacterium* species involves the EMP pathway, the TCA cycle and several anaplerotic mechanisms (see Fig. 9.1 on p. 152).

Some of the major anaerobic metabolic pathways and the end-products formed are summarized in Fig. 2.10. Pyruvate is produced via the EMP pathway in important anaerobic fermentations involving strains of *Saccharomyces*, *Lactobacillus* and *Clostridium* species. In yeast alcoholic fermentations, the pyruvate is converted to acetaldehyde and then to ethanol. $NADH + H^+$ formed during glycolysis is re-oxidized by the enzyme alcohol dehydrogenase during conversion of acetaldehyde to ethanol. Addition of sodium bisulphite to the yeast alcoholic fermentation results in the formation of an acetaldehyde–sulphite complex and the reduction of the glycolytic intermediate dihydroxyacetone phosphate to glycerol phosphate, leading to glycerol production (see Fig. 8.4 on p. 140). In the production of L-(+)-lactic acid by industrial fermentation using *Lactobacillus delbruckii*, the pyruvate formed from hexose is converted to lactate by the enzyme L-lactate de-hydrogenase. While acetic acid can be produced biologically by aerobic conversion of ethanol to acetate (see Chapter 7) the overall stoichiometric yield from glucose is 0.66. *Clostridium thermoaceticum* ferments 1 mole of glucose stoichiometrically to 3 moles of acetate via pyruvate (see Fig. 8.5 on p. 141). *Clostridium acetobutylicum* converts pyruvate to mixtures of acetone and butanol. Anaerobic digestion in waste treatment processes involves the interaction of three microbial groups in a complex series of reactions. Organic acids, aldehydes and

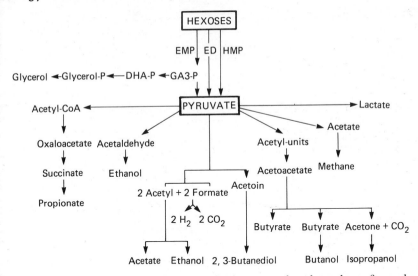

**Fig. 2.10** Major anaerobic metabolic pathways and end-products formed.

esters produced by degradation of polymeric material and monomers are converted to acetate, $CO_2$ and hydrogen. Methane and $CO_2$ are then produced from the acetate and hydrogen (see Chapter 12).

More detailed aspects of the primary metabolism of microbial strains involved in specific industrial fermentation processes are discussed in some of the chapters which follow.

SECONDARY METABOLISM

Secondary metabolites are compounds which are not required for cellular biosynthesis and which do not play a direct role in energy metabolism. Consequently, the reason that secondary metabolites exist is unclear. Some secondary metabolites, namely antibiotics, are advantageous to the organism in that they inhibit other organisms in their surroundings from competing for available nutrients. Others may act as effectors for differentiation, agents of symbiosis or factors in sexual cycles. They tend to be synthesized as families of related compounds.

Secondary metabolite biosynthesis is closely interrelated to the primary metabolism of the producing cells. Precursors of secondary metabolites are usually normal or modified primary metabolites such as organic acids, isoprene units, aliphatic and aromatic amino acids, sugar derivatives, cyclitol derivatives and purine or pyrimidine bases. The most important group of microbial secondary metabolites produced by industrial fermentation are the antibiotics (see Chapter 10). Other examples of commercial microbial secondary metabolites include gibberellic acid, ergot alkaloids, antitumour substances, immunomodulators and insecticides. A number of high-value secondary metabolites are extracted from plant material. Examples include the pharmaceuticals vinblastine, vincristine, ajmalicine, digitalis and codeine, the flavour or fragrance compounds quinine, jasmine and spearmint and the pyrethrin insecticides. It is possible that some of these will be produced by suspension culture in the future. A certain degree of cell aggregation and cell differentiation seems to be necessary to obtain higher yields of plant secondary metabolites. For large-scale plant-cell culture this is not always desirable because generally the culture period for differentiation is rather longer than that of undifferentiated cells which adds to costs and increases risks of microbial contamination.

## Regulation of metabolism

A particular cell has the genetic potential to form over 1000 enzymes. Clearly with such a complex network of metabolic pathways operating in the cell, these enzymes must be synthesized in the correct proportions and must operate in a co-ordinated manner to enable the cell to operate as an efficient unit. Micro-organisms are able to alter their composition and metabolism in response to environmental changes. The major regulatory mechanisms which confer this flexibility operate at both the level of enzyme synthesis and at the level of enzyme action.

In a given fermentation the flux of substrate through the complex network of catabolic and anabolic pathways will be influenced by a variety of mechanisms. Selected rate-limiting enzymes are strategically placed in the various pathways and these key enzymes are usually sensitive to induction/repression and activation/inhibition mechanisms (see below). Over-production of primary metabolites results from high differential flux rates on the upstream side leading to accumulation of the metabolite and reduced or negligible flux rates downstream which would cause metabolite degradation or utilization. The processes regulating secondary metabolite production are more complex and less well understood but there is evidence that many of the mechanisms are similar to those controlling primary metabolism.

Clearly, in the development of fermentation processes for over-production of a particular metabolite, it is essential that account is taken of the various regulatory processes, in the design of optimized environmental conditions and in the isolation or manipulation of microbial production strains.

ENZYME SYNTHESIS AND DEGRADATION

The total concentration of a particular intracellular enzyme present in cells is regulated by the relative rates of synthesis and degradation or inactivation of that enzyme. Proteins in cells are continually being degraded by proteolytic enzymes, such that each protein has an average limited half-life. Proteins also vary in their susceptibility to proteolytic attack. Net synthesis of an enzyme occurs when the rate of synthesis is greater than the rate of degradation.

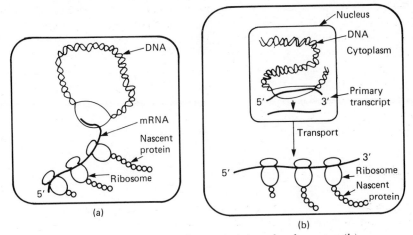

**Fig. 2.11**   Protein synthesis in prokaryotes (a) and eukaryotes (b).

The general mechanism of protein synthesis in prokaryotes is illustrated in Fig. 2.11a. A single RNA polymerase recognizes and binds to specific DNA sites called promoters, initiating mRNA synthesis and transcription of an operon. Beyond the end of the structural gene a termination region causes RNA polymerase to cease transcription and dissociate from the DNA. In prokaryotes many mRNAs are polycistronic, that is, they encode multiple polypeptide chains (almost all eukaryotic mRNAs are monocistronic). Synthesis of each protein is initiated by the binding of ribosomes to a specific site on the mRNA and translation starts even before the RNA polymerase has reached any signal for termination.

In some cases bacterial transcription is initiated at a constant (usually slow) rate without regulation. In other cases, the rate at which RNA synthesis is initiated is governed by proteins that interact in a site specific manner with DNA regions termed operators, which are located near promoter sites. The regulatory proteins that bind to operators are called repressors or activators depending on whether they decrease or increase the rate of transcription initiation. Some operons are controlled by both activator and repressor proteins and in prokaryotes the promoter–operator sites often control the transcription of DNA encoding for more than one protein. Most regulation of protein synthesis takes place through control of the rate of initiation of transcription. However in some cases regulation takes place at the level of transcription termination or at translation initiation sites.

The mechanism of protein synthesis in eukaryotes is illustrated in Figure 2.11b. Transcription occurs in the nucleus and the primary RNA transcript is extensively modified in the nucleus before it emerges into the cytoplasm to associate with ribosomes. Control of protein synthesis in eukaryotes may be exerted at the level of RNA processing as well as at transcription or translation. There is very little translational regulation in prokaryotes.

In both prokaryotes and eukaryotes the stability or half-life of mRNA will also influence the rate of protein synthesis. Different mRNAs have different stabilities so that the effect on overall protein synthesis is complex.

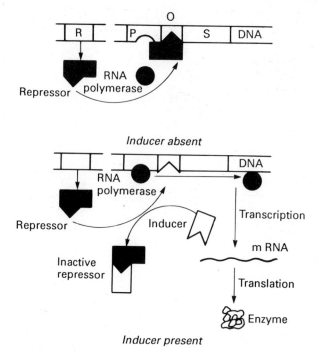

**Fig. 2.12** Diagrammatic representation of an inducible enzyme system.

*Induction*

Some enzymes are produced constitutively, that is, they are always made in substantial amounts when cells are growing. Other enzymes are inducible, that is, they are formed only in the presence of key substrates in the medium. The structural genes of inducible enzymes are normally inactive or operate at a very low 'basal' rate in the absence of substrate. When the substrate is present, the structural gene is switched on and transcription and translation occur. This process is illustrated using the Jacob and Monod model for induction of $\beta$-galactosidase by *E. coli* (Fig. 2.12). A regulator gene synthesizes a repressor which binds to an operator gene in the absence of inducer, preventing RNA polymerase from moving along the DNA to transcribe the structural gene. The inducer binds to the repressor, removing it from the operator gene and allowing transcription to take place.

*Catabolite repression*

This phenomenon frequently occurs when the cell is grown in a medium containing more than one utilizable growth substrate. In carbon catabolite repression, enzymes are synthesized which catabolize the best substrate, usually glucose, and only after exhaustion of this substrate are enzymes produced to break down the poorer growth substrate. Inhibition of cyclic 3',5'-adenosine monophosphate (cAMP) formation appears to be the key factor in catabolite repression in *E.*

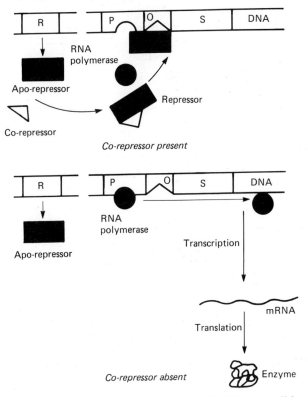

**Fig. 2.13** Diagrammatic representation of a feedback repressible enzyme system.

*coli.* cAMP is necessary for synthesis of mRNA. Glucose metabolism reduces the cAMP content of cells by 1000-fold. cAMP level is a function of adenyl cyclase and cAMP phosphodiesterase which make and degrade cAMP, respectively. Adenyl cyclase appears to be inhibited by transport of readily utilizable carbon sources into the cell. While glucose catabolite repression is widely observed in fungi, yeasts and *Bacillus* species, the regulatory mechanisms are not well understood.

Many enzymes, including proteases are repressed by rapidly utilizable amino acids or ammonia (nitrogen catabolite repression). Limitation of ammonia in fermentation growth media usually de-represses synthesis of these enzymes.

*Feedback repression*
While synthesis of catabolic enzymes is often controlled by induction and catabolite repression, biosynthesis of anabolic enzymes may be regulated by feedback or end-product repression. Feedback repression is widely used in nature to control amino acid, purine, pyrimidine and vitamin synthesis. In the classical model explaining feedback repression (Figure 2.13) the regulator gene produces an apo-repressor protein, which when combined with a co-repressor (end-product), binds to the operator site and blocks transcription. In the absence of the end-product, no repression takes place.

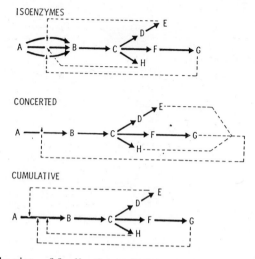

**Fig. 2.14**   Mechanism of feedback regulation of branched metabolic pathways (reproduced with permission from Demain, 1971).

## MODERATION OF ENZYME ACTIVITY

Many enzymes have single or multiple sites other than their active site, called allosteric sites, which reversibly bind particular metabolites or effector molecules, causing a change in enzyme conformation and thereby increasing or decreasing its activity. The activities of other enzymes are regulated by covalent modulation, whereby enzymatic formation or degradation of a covalent bond in the enzyme serves to activate or inactivate enzyme activity. Compounds such as ATP, ADP, AMP and nicotinamide nucleotides, which reflect the level of the cell's available energy, also modulate enzyme activity and control metabolism by adjusting the balance of catabolic processes and energy-using anabolic processes.

## REGULATION OF BRANCHED METABOLIC PATHWAYS

Biosynthetic metabolic pathways often have a common enzyme sequence and then branch leading to more than one end-product. Micro-organisms have evolved feedback mechanisms, whereby a build-up of one end-product causes a feedback effect on the first enzyme of the branch leading to that product. In addition, mechanisms exist whereby the end-product of a branched pathway causes partial feedback inhibition of the first enzyme of the common sequence so that the flux of substrate passing through this sequence is proportionately reduced. This effect is achieved by use of isoenzymes, concerted feedback regulation and cumulative feedback regulation (Fig. 2.14). These regulatory effects can be of two types: inhibition of enzyme activity and repression of enzyme synthesis. Where isoenzymes (multiple enzyme forms capable of catalysing the same reaction) are

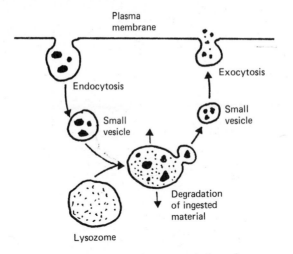

**Fig. 2.15**  Processes of endocytosis and exocytosis in eukaryotes.

involved, the synthesis or inhibition of each enzyme form may be regulated by a different end-product. With concerted feedback regulation, only one enzyme is involved, but more than one product must be present to inhibit activity or repress enzyme synthesis. With cumulative feedback regulation, each end-product causes partial inhibition or repression and all end-products are required to completely block activity or synthesis.

## Substrate assimilation/product secretion

Efficient substrate influx and product efflux mechanisms are essential for rapid cell growth, supply of substrate for conversion to primary and secondary metabolites and other products and for product excretion. In prokaryotes, substrates move into and out from the cell directly by traversing membranes associated with the cell envelope. Some bacterial fermentations have been manipulated to alter the cell membrane or wall composition in order to modify uptake mechanisms or facilitate secretion (see Chapter 9: glutamic acid production). In addition to having an external plasma membrane, eukaryotes also contain an extensive system of internal membranes and organelles. Some materials are transported into the cell by endocytosis. In this process a region of the plasma membrane invaginates and forms a vesicle containing material from outside the cell. The vesicle fuses with another organelle, a primary lysozome, to form a secondary lysozome where much of the external material can be degraded and transported to the cytoplasm. Other degradation products are packaged in vesicles and subjected to exocytosis (Fig. 2.15). Such vesicles are present in large numbers at the hyphal apices in fungi. Consequently, in eukaryotes many of the processes for transport of substrates and products across membranes occur

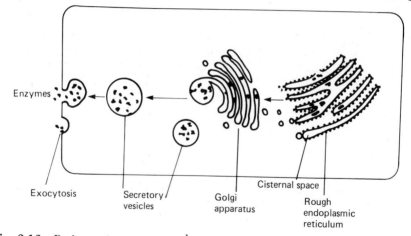

**Fig. 2.16**  Pathway for secretion of proteins in eukaryotes.

between cell protoplasm and organelles such as vesicles and the endoplasmic reticulum.

Solutes are transported across membranes without modification (a) by passive diffusion down a concentration gradient, (b) by facilitated diffusion down a concentration gradient, involving a carrier system, which, like enzymes, displays saturation kinetics, and (c) by active transport, which also involves a carrier system but allows net movement of the solute against an electrochemical gradient. Solutes may also be modified during translocation by enzymes with vectorial properties.

Specific transport proteins appear to be responsible for transport of many solutes across membranes, some of which are energy-coupled and some of which are inducible. In the biosynthesis of some microbial extracellular products, the final biosynthetic enzyme may be membrane-associated and function to facilitate excretion. For example, translocases have been implicated in the extrusion of exopolysaccharides from cells.

ENZYME SECRETION

Gram-positive prokaryotes and eukaryotes have a single plasma membrane at their surface. In contrast, Gram-negative bacteria have cell walls containing two membranes separated by a periplasmic space. Extracellular enzymes, synthesized by Gram-positive bacteria, are transported through the cytoplasmic membrane, diffuse through the cell wall and accumulate in the extracellular culture fluid. Extracellular enzymes or secretory proteins are synthesized by eukaryotes at the surface of the endoplasmic reticulum (ER) and transported through the membrane into the ER cisternal space. The protein contained in the cisternal spaces is later processed through the Golgi apparatus and other vesicles and

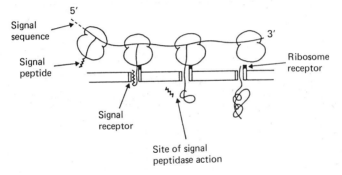

**Fig. 2.17**  Diagrammatic illustration of co-translational secretion (reproduced with permission from Priest, 1984).

secreted into the cell's external environment by fusion of the vesicles with the plasma membrane (Fig. 2.16). Glycosylation of glycoproteins occurs in the ER cisternal space and in the Golgi bodies. This general secretory pathway appears to be similar in higher eukaryotes and in yeast and probably exists in filamentous fungi as well, although variations in the process may exist. In Gram-negative bacteria, many enzymes are transported across the cytoplasmic membrane and are located in the periplasmic space.

Proteins which are secreted across membranes usually contain an amino terminal peptide extension, called the signal sequence. The signal peptide, emerging from the ribosome, interacts with the inner surface of the plasma membrane of bacteria or the ER of eukaryotes, forming a pore or channel. As the polypeptide elongates it passes through the pore in the membrane, in a process known as co-translational secretion, and a signal peptidase located on the external side of the membrane removes the signal peptide, enabling the protein to assume its normal three-dimensional structure. The signal hypothesis for co-translational protein secretion is illustrated diagrammatically in Fig. 2.17. Some extracellular proteins are secreted after they have been synthesized in a process (also involving a signal peptide) known as post-translational secretion.

# Chapter 3

# *Fermentation systems*

## Batch and continuous processes

Fermentations may be carried out as batch, fed-batch or continuous processes or variations of these procedures. Batch culture may be considered as a closed system (except for aeration) containing a limited amount of medium, in which the inoculated culture passes through a number of phases as illustrated by a typical batch growth curve (see Fig. 2.6 on p. 23). Whereas in batch culture all the substrate is added at the beginning of the fermentation, in fed-batch processes substrate is added in increments throughout the process. Fed-batch cultures may be operated to remove the repressive effects of rapidly utilized carbon sources, to reduce the viscosity of the medium, to reduce the effect of toxic medium constituents or simply to extend the product formation stage of the process for as long as possible. Batch or fed-batch culture systems are used in the majority of industrial fermentation processes and are particularly suited to fermentations where the bulk of product formation occurs after the exponential growth phase. The major disadvantage of batch-type fermentations, used for production of growth-associated products, is that efficient product formation only occurs during a fraction of each fermentation cycle (Fig. 3.1). Continuous systems with continuous high output can consequently be much more efficient in terms of fermenter productivity (product output per unit volume per unit time, $kg\,m^{-3}\,h^{-1}$) for certain applications. Continuous fermentations may be considered as open systems in which medium is continuously added to the bioreactor and an equal volume of fermented medium is simultaneously removed. There are two main types of continuous reactors, homogeneously mixed reactors and plug flow reactors.

The biomass concentration at any time (t) in a batch culture is

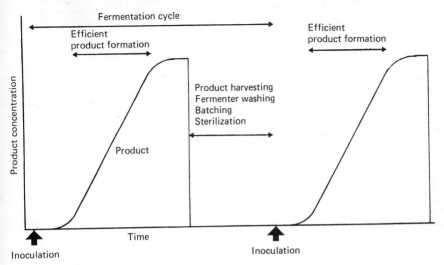

**Fig. 3.1** Batch-type fermentation cycle.

$$x - X_R = \Upsilon(S_R - s)$$

Where $x$ = the cell concentration at time t

$X_R$ = the inoculum or initial cell concentration

$\Upsilon$ = the yield factor for the limiting substrate (grams of biomass per gram of substrate consumed)

$s$ = the substrate concentration at time t

$S_R$ = the original concentration in the medium

Therefore

$$\Upsilon = \frac{x - X_R}{S_R - s}$$

The amount of biomass attained in the stationary phase is ideally dependent on the original concentration of the substrate which limits growth and the efficiency of the organism in converting the substrate into cellular material. Therefore assuming the growth limiting substrate S = 0 in the stationary phase

$$\Upsilon = \frac{x - X_R}{S_R}$$

In a homogeneously-mixed single stage continuous reactor, addition of substrate to the reactor at a suitable rate with displacement at an equal volumetric rate of culture from the vessel produces a steady state where formation of new biomass by the culture is balanced by the loss of cells from the culture. Under these conditions:

(a) the specific growth rate ($\mu$) is controlled by the dilution rate ($D$) since $\mu = D$

The dilution rate $D$ (h$^{-1}$) representing the flow of medium through a chemostat system per unit volume of the contained liquid is defined as

$$D = F/V$$

where $V$ = the volume (m$^3$)

$F$ = the flow rate (m$^3$ h$^{-1}$)

The net change in biomass over time assuming all cells are viable, may be expressed as

$$\mathrm{d}x/\mathrm{d}t = \text{growth} - \text{output}$$
$$= \mu x - Dx$$

Under steady-state conditions, reactor cell concentration is constant, therefore

$$\mathrm{d}x/\mathrm{d}t = 0$$
$$\mu x = Dx$$
$$\mu = D$$

When $D \leqslant 0.9\mu_{max}$, then the rate of cell production in the fermenter is exactly balanced by the rate of cell output, i.e. a steady-state condition ($\bar{x}$ = constant) is established. On the other hand, when $D$ approaches $\mu_{max}$ in value, the output of cells in the effluent exceeds their rate of production by growth in the fementer and 'washout' occurs i.e. $\mathrm{d}x/\mathrm{d}t$ is negative

But $\mu$ is related to substrate concentration ($s$) by the Monod equation

$$\mu = \frac{\mu_{max}s}{K_s + s}$$

At steady-state with $\mu = D$, and for the Monod kinetic model of cell growth

$$D = \frac{\mu_{max}\bar{s}}{K_s + \bar{s}}$$

where $\bar{s}$ is the steady-state residual substrate concentration

$$\bar{s} = \frac{K_s D}{\mu_{max} - D}$$

As $D \to \mu_{max}$, then $\bar{s} \to S_R$

The biomass concentration in the chemostat at steady-state for the case of sterile feed ($X_R$ or $X_F = 0$) is given by

$$\bar{x} = Y(S_R - \bar{s})$$

Therefore

$$\bar{x} = Y\left[S_R - \frac{K_s D}{\mu_{max} - D}\right]$$

As $D \to \mu_{max}$, $\bar{x} \to 0$

The biomass productivity, $P'_c$, for continuous, steady-rate fermentation is given by

$$P'_c = D\bar{x}$$

**Fig. 3.2**   Kinetics of continuous culture.

(b) the steady-state substrate concentration $\bar{s}$ in the chemostat is determined by the dilution rate according to the equation

$$\bar{s} = \frac{K_s D}{\mu_{max} - D}$$

if kinetics are characterized by the Monod model and as long as $D < \mu_{max}$

(c) the steady-state biomass concentration $\bar{x}$ is determined by the operational variables $S_R$ and $D$ according to the equation

$$\bar{x} = Y\left[ S_R - \frac{K_s D}{\mu_{max} - D} \right]$$

These equations are derived in Fig. 3.2.

The effect of dilution rate on steady-state biomass and substrate concentrations and biomass productivity is illustrated in Fig. 3.3. As $D$ approaches $\mu_{max}$, the concentration of residual limiting substrate required to support the increased growth rate rises and thus residual substrate concentration increases (to the upper limit $s = S_F$) and biomass decreases. When a continuous culture system is used, as in industrial fermentation for SCP production, the quantity of cells produced per unit time (the output rate) and the quantity of cells produced per unit weight of substrate utilized ($Y$) must be maximized. Under steady-state conditions, output rate will be the product of flow rate and cell concentration. For maximum cell output, dilution rate should be high but cannot exceed $\mu_{max}$ or washout will occur. Maximum productivity, combining a high output with efficient substrate utilization, will be obtained with a flow rate at, or slightly below, the maximum output rate and with the highest practical substrate concentration in the feed.

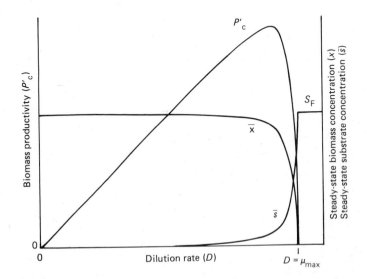

**Fig. 3.3** Effect of dilution rate on steady-state biomass ($\bar{x}$) and substrate ($\bar{s}$) concentrations and biomass productivity ($P'_c$).

A similar approach may be adopted to optimize chemostat conditions for production of some primary metabolites. However, in the case of secondary metabolites where the product is not growth-associated, a single optimally-mixed chemostat may not be used. Multi-stage chemostats may be used to enable different environmental conditions to prevail in the separate stages, in situations where optimal conditions for growth and product formation are different. Chemostats incorporating concentrated biomass recycle processes to increase reactor biomass concentration have found industrial applications in alcohol production and effluent treatment.

✦ In plug flow reactors, the culture flows through a tubular reactor without back-mixing. At the inlet of the reactor, cells and medium are continually added. The composition of the medium, biomass concentration and product concentrations vary along the length of the reactor in a manner analogous to the changes occurring with time in a batch fermenter. A biomass recycle process usually must be incorporated in order to provide a continuous inoculum at the inlet.

## Fermenter design

The main function of a fermenter is to provide a controlled environment which allows for efficient growth of cells and product formation. The modern standard laboratory fermenter is designed as a sophisticated unit having the properties and instrumentation necessary to develop and operate a variety of fermentation processes. In discussing the features of pilot fermenters, it should be remembered that industrial production-scale fermenters are usually purpose-built units, cost-effectively designed for a specific process or collection of related processes. Consequently, they only contain the design and instrumentation features specified for the particular production process.

REQUIREMENTS FOR ASEPTIC OPERATION

Many bioreactor systems have to operate under aseptic conditions using a pure culture inoculum of the process micro-organism and excluding unwanted contaminating organisms. Consequently, the bioreactor and associated pipework is designed as a pressure vessel such that the system and its medium contents can be sterilized at the appropriate temperature/pressure, a minimum of 121°C/15 psi for 15–30 min. Valves used in sterilizable sections must be suitable for aseptic operation. Pinch valves and diaphragm valves are particularly suitable as the operating mechanism is separated from the medium liquid or gas by a flexible tube or diaphragm. All entry points to the vessel and the processes for adding and removing gases or liquids to/from the vessel during the fermentation are designed to maintain aseptic conditions (Table 3.1). An aseptic sampling port is shown in Fig. 3.4.

**Table 3.1** Criteria in fermenter design and operation related to sterility maintenance

| *Component or process* | *Principle* | *Example or comment* |
|---|---|---|
| 1. Vessel | Designed for steam-sterilization under pressure, to withstand chemical corrosion and be non-toxic to the organism | (a) Glass of appropriate thickness for small fermenter pots and for fermenter sightglass <br> (b) Stainless steel resistant to defined pH |
| 2. Entry and exit points to fermenter | All steam-sterilizable | |
| 3. Pipework | No pockets or dead spaces, slope to drainage point | |
| 4. Valves | Must be suitable for aseptic operation. Materials of construction should be able to tolerate the temperature and pressure conditions of the process and be resistant to chemicals being used | Valves vary in their degree of suitability. Pinch valves and diaphragm valves are particularly suited to aseptic use because the valve mechanism is separated from the flowing broth or gas by a flexible tube and flexible diaphragm, respectively |
| 5. Impeller glands and bearings | Seals to be designed for prolonged aseptic operation | (a) Packed gland—layers of asbestos or cotton yarn are packed against the shaft <br> (b) Seals: Bush seals—aseptic conditions maintained by precision flat surface contact between a rotating bush seal on the agitator shaft and a stationary bush seal on the shaft. With mechanical seals, the stationary component and the rotating component are pressed together by springs or expanding bellows <br> (c) Magnetic drives—shaft does not penetrate vessel. An external drive shaft rotates an internal impeller shaft by magnetic forces |

**Table 3.1**  continued

| Component or process | Principle | Example or comment |
|---|---|---|
| 6. Gas inlet | Air filter to sterilize inlet air. Should be steam-sterilizable. Non-return valve located between sparger and air filter to prevent medium flow back to filter | (a) Glass wool, glass fibre or other packing to physically trap particles<br>(b) Pleated membrane filter made of cellulose ester, polysulphone nylon or other material |
| 7. Gas outlet | Design such that back flow contamination does not occur | Outlet filter may be used where release of culture strains to the environment is undesirable |
| 8. Feed additives | Feeds sterilized by heat or filtration as required | Acid, alkali (liquid or gaseous), carbohydrate, precursor |
| 9. Feed, inoculum and sampling lines | Steam-sterilizable before and after use | Sampling sequence (see Fig. 3.4): (a) open valves 2 and 3 to sterilize piping, (b) close valve 3, (c) open valve 4 to cool hot sterile pipe before sample is taken, (d) close valve 4, open valve 1, (e) discard dead space broth, (f) collect sample, (g) close valve 1, (h) repeat (a) and (b) |

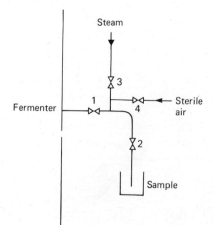

**Fig. 3.4**  Aseptic sampling port. 1, 2 and 3 are valves.

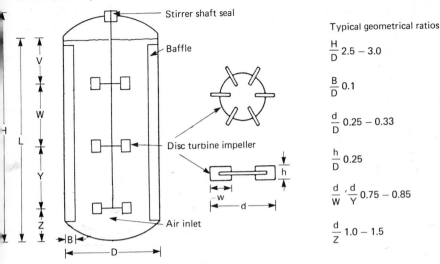

**Fig. 3.5** Diagram of fermenter and flat-bladed disc turbine impeller

Not all fermentation processes require absolutely sterile conditions. Some processes simply reduce contamination by pasteurization while other processes actually depend on the indigenous or natural microbial population. The reactor design requirements used for these processes may be simplified accordingly.

## REQUIREMENTS FOR AERATION AND MIXING

Aerobic industrial fermentation processes require a system for aeration and mixing of the culture. The conventional stirred-tank reactor contains a mechanical agitation system involving a vertical shaft with a number of impellers. In most bioreactors flat-bladed (Rushton) disc turbines are used and are located along the shaft and have characteristic dimensions to achieve good mixing and dispersion throughout the vessel medium. A number of baffles (usually 4–6) having a width of 10% of the vessel diameter are located at the vessel perimeter to increase turbulence. Sterile air is introduced at the base of the tank, usually through a ring sparger, and effectively dispersed throughout the medium by the agitation system. Important features of the agitation system are summarized in Fig. 3.5.

The geometry of aerated fermenters is designed to facilitate efficient gas exchange. The actual specification required will take account of transport phenomena in bioprocess systems.

### Fluid flow
Fluids are described as Newtonian when their flow characteristics obey Newton's law of viscosity. When a fluid is contained between two parallel plates of area $A$ and distance $x$ apart and if one plate is moved in one direction at a constant velocity, the liquid 'layer' adjacent to the moving plate will move in that direction

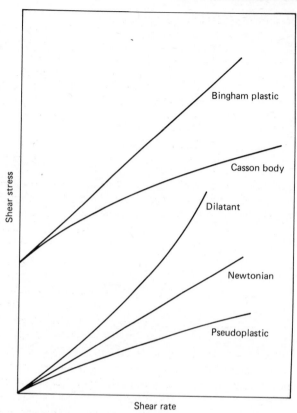

**Fig. 3.6**   Rheograms of fluids having different rheological properties.

and impart some of its momentum to the next layer, causing it to move at a slightly slower rate. Newton's law of viscous flow states that the viscous force $F$ opposing motion at the interface between the two liquid layers, flowing with a velocity gradient of $dv/dx$ is described by the equation

$$F = \eta A \, dv/dx$$

and

$$\eta = \frac{(F/A)}{dv/dx}$$

where    $\eta$ = the fluid viscosity or the fluid's resistance to flow
         $F/A$ = the shear stress or force per unit area
         $dx/dx$ = the shear rate or the velocity gradient

Viscosity is the ratio of shear stress to shear rate and a rheogram plot of shear stress against shear rate produces a straight line for a Newtonian fluid with a constant viscosity equal to the slope of the line. Whereas the viscosity of a Newtonian fermentation broth will not vary with shear or agitation rate, the viscosity of a non-Newtonian fluid will vary with shear or agitation rate. The

**Table 3.2**  Rheology of mycelium-producing cultures

| *Micro-organism* | *Product* | *Culture rheology* |
|---|---|---|
| *Coniothyrium hellbori* | Steroid hydroxylation | Bingham |
| *Endomyces* sp. | Glucoamylase | Pseudoplastic |
| *Penicillium chrysogenum* | Penicillin | Pseudoplastic |
| *Streptomyces griseus* | Streptomycin | Bingham |
| *Streptomyces kanamyceticus* | Kanamycin | Bingham |
| *Streptomyces niveus* | Novobiocin | Bingham |

rheogram plots of a Newtonian and some non-Newtonian fluids are illustrated in Fig. 3.6. Most polymer solutions behave as pseudoplastic fluids which are characterized by having a decreasing apparent viscosity with increasing shear or agitation rate. Some fermentation broths have been characterized as having pseudoplastic or Bingham plastic rheology (Table 3.2).

*Mass transfer*
During microbial growth and metabolism, the continuous transfer of nutrients and metabolites occurs between the external environment and the cell. During this exchange, a nutrient or metabolite passes through several phases of resistance, which may be in the form of solids, liquids or gases. In conventional fermentation processes, microbial demand for substrates other than oxygen is usually met without difficulty, as these are supplied in excess in solution in the medium.

The rate of mass transfer between the gas and liquid phase is strongly influenced by the solubility of the gas in the liquid phase. Oxygen is only sparingly soluble in aqueous solutions. Consequently, in many aerobic fermentation processes, supply of oxygen becomes the crucial factor determining metabolic rates and oxygen mass transfer has an important bearing on fermenter design.

*Oxygen transfer*
The transfer of oxygen to the cell during fermenter aeration involves oxygen transfer from air bubbles into solution, transfer of oxygen through the medium to the microbial cell and finally uptake of oxygen by the cell. Oxygen transfer from air bubbles into solution which is the rate-limiting step for non-viscous fermentations, and is even more limiting for viscous fermentation broths, may be described by the equation

$$\frac{dC_L}{dt} = K_L a (C^* - C_L)$$

where $C_L$ is the dissolved oxygen concentration in the bulk of the liquid $(mmol\,dm^{-3})$
  $t$ is time (h)
  $dC_L/dt$ is the change in oxygen concentration with time (mmoles $dm^{-3}$ $h^{-1}$) or the oxygen-transfer rate (OTR)
  $K_L$ is the mass-transfer coefficient $(cm\,h^{-1})$

$a$ is the gas–liquid interface area per liquid volume $(cm^2 cm^{-3})$
$C^*$ is the saturated dissolved oxygen concentration $(mmol dm^{-3})$. At 1 atm at 30°C, solubility of $O_2$ in water is 1.16 mmol $dm^{-3}$.

$K_L a$, the volumetric mass-transfer coefficient $(h^{-1})$, is a measure of the aeration capacity of the fermenter under test conditions—the larger the $K_L a$, the higher the aeration capacity. The dissolved-oxygen concentration in a fermenter medium will be a balance between the supply of dissolved oxygen and the demand by the organism. If the $K_L a$ of the fermenter is too low to supply the oxygen demand of the organism, dissolved oxygen concentrations will fall below the critical level $(C_{crit})$, usually less than 0.05 mmol $dm^{-3}$. The $K_L a$ of the fermenter must therefore be large enough to maintain the optimum dissolved-oxygen concentration for product formation. Factors which affect the $K_L a$ value achieved in a fermentation vessel of specified geometry include air-flow rate, rate of agitation, (ie impeller rotational speed) broth physicochemical and rheological properties and presence of antifoam agents. When a fermentation is scaled-up, it is important that the optimum $K_L a$ value found at the lower scale is employed at the larger scale. Thus $K_L a$ values and knowledge of the factors affecting $K_L a$ are important in fermentation design and scale-up. Particularly with non-Newtonian fermentations the scale-up strategy attempts to maintain constant tip speed as well as maintaining a constant $K_L a$ by varying the impeller diameter to vessel diameter ratio in the larger vessel.

*Energy transfer*

For growth, metabolism and mass transfer to occur at the desired rates, energy has to be supplied or removed from the reactor system. Heat energy is exchanged during sterilization and process temperature control. Metabolic energy is produced during metabolism and dissipated as heat which tends to increase the temperature of the fermentation broth. Some heat will also be generated as a result of the mechanical energy input from mechanical agitation and the aeration processes. When bioreactors are run at temperatures above ambient, a source of heat is required to maintain the temperature. The net requirement for external heating or cooling of the fermenter will be determined by the balance of these heat-generation and heat-loss processes and usually require heat-exchange systems. With fermenters, this heat exchange is achieved by water circulation through a jacket on the outside of the vessel or through coils located inside the vessels. Large aerobic fermentations are usually exothermic during active growth and metabolism and require cooling for temperature control.

*Fermenter power requirements*

Efficient mixing in baffled agitated fermenters is achieved by using fully developed turbulent flow which can be characterized by the dimensionless Reynolds number $(N_{Re})$

$$N_{Re} = \frac{D^2 N \rho}{\eta} \quad (> 10^5 \text{ for Rushton turbines})$$

where $D$ = the impeller diameter (cm)

$N$ = the impeller rotational speed $(s^{-1})$

$\rho$ = the liquid density $(g\, cm^{-2})$

$\eta$ = the fluid viscosity

When a non-gassed Newtonian liquid is agitated in a baffled vessel, power input to the liquid can be represented by another dimensionless number termed the power number $(N_p)$.

$$N_p = \frac{P}{\rho N^3 D^5}$$

where $P$ is the external power for the agitator (non-gassed).

In a fully baffled, agitated fermenter, the power number $(N_p)$ is related to the Reynolds number by the equation

$$N_p = C(N_{Re})^z$$

where $C$ is a constant, dependent on vessel geometry but independent of vessel size and $x$ is an exponent. Therefore

$$\frac{P}{\rho N^3 D^5} = C\left[\frac{(D^2 N\rho)^z}{\eta}\right]$$

so that values for $P$ at various values of $N$, $D$, $\eta$ and $\rho$ may be determined experimentally.

Typical power requirements for fermenters of capacities 100 litres, 2500 litres and 50 000 litres would be 1–2 kW, 15 kW and 100 kW, respectively.

## OTHER FERMENTER DESIGNS

### Submerged culture systems

A modified aerated STR, the Frings acetator, is used in vinegar production. In this design, air is drawn in and distributed via a high-speed hollow-body turbine rotor, connected to an air suction pipe. The aerator is self-aspirating and so compressed air is not required (see Fig. 7.8 on p. 127).

The STR is the most widely used bioreactor for aerated fermentations perhaps because of its reliability and flexibility. Nevertheless, operating and investment costs are relatively high. In addition, problems exist in designing impellers for very large fermenters due to the length of the impeller shaft. In bioreactors without mechanical agitation, such as tower and loop bioreactors, aeration and mixing is achieved with high gas through-puts.

In tower fermenters (Fig. 3.7), air is introduced at the base of the fermenter and mixing is due to the rising bubbles so that the shear on the organism is minimal. Tower fermenters have been used to produce citric acid, by pellets of *A. niger* and by strains of *Candida guilliermondii*, and also for production of vinegar, industrial alcohol, and beer. Because vertical mixing is relatively poor in tower fermenters, they can be operated as continuous systems, with bottom entry feed and top exit,

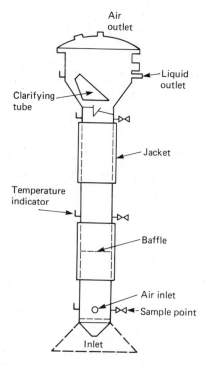

**Fig. 3.7** Tower fermenter (reproduced with permission from Kristiansen and Chamberlain, 1983).

with high biomass retention. In non-aerated tower fermenters, used for brewing or industrial alcohol production, biomass retention is maximized by use of fermenting yeasts having good flocculating properties.

The loop air-lift bioreactor contains either an internal or external draft tube, sometimes baffled, which increases mixing by forcing a directional flow of the bulk liquid (Fig. 3.8). The driving force for circulation is created by the difference in density (due to differences in the amount of the dispersed air bubbles) between the riser and down flow sections. In the ICI pressure cycle reactor, for example, air is introduced at the base of the fermenter and forced into solution by the hydrostatic pressure of the column. (See Fig. 6.1d and 6.1e on p. 97). It will be seen in Chapter 6 that designs such as these have been used successfully for production of biomass by unicellular organisms. For cultivation of viscous filamentous mycelial cultures for SCP production combined draft tube–impeller systems have been developed (see Fig. 6.1f). In industrial fermentations involving animal and plant cells, which are more susceptible to shear damage than microbial cells, advantage has been taken of the lower shear effect of air-lift fermenters.

Variations of these submerged culture designs are used in activated sludge waste-treatment systems. The conventional process involves vigorous agitation with air or oxygen injected by bubble diffusers, paddles, stirrers, etc. (Fig. 3.9).

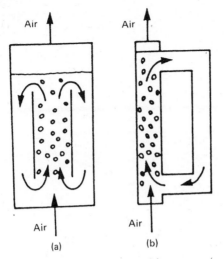

**Fig. 3.8** Internal (a) and external (b) loop bioreactor (reproduced with permission from Smith, 1985).

Deep shaft systems have also been developed based on the principles of the loop air-lift fermenter (Fig. 3.10).

Recent dramatic progress in animal-cell culture and hybridoma technology has lead to the development of novel industrial culture processes for production of animal-cell products. The simplest way to grow mammalian cells is in suspension cultures. Most cells of leukemic or lymphoid origin grow as single cells in either stationary or agitated suspension cultures. Many anchorage-dependent cells will not grow in suspension culture unless special procedures are used. The most promising method in such cases involves use of microcarrier beads, made of natural or synthetic polymers, as a support for anchorage-dependent cells. Problems associated with large-scale production of animal cells in suspension culture include the limited cell densities achievable in conventional cell-culture

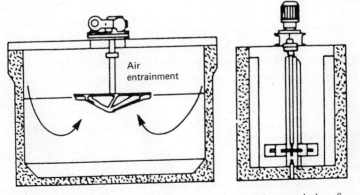

**Fig. 3.9** Activated sludge bioreactor (reproduced with permission from Smith, 1985).

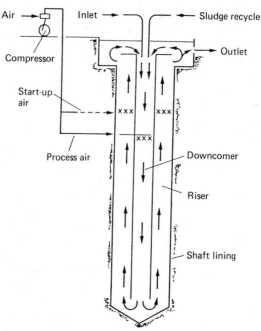

**Fig. 3.10** Deep-shaft effluent treatment plant (reproduced with permission from Wheatley, 1984).

systems and cell sensitivity to shear. Gently stirred and air-lift fermenters have been used as well as more novel reactors. Encapsulation of hybridoma cells in alginate spheres allows stirred tank reactors to be used and produces high concentrations of cells and antibodies within the capsule. Another system involves use of perfusion systems, involving porous tubes and semi-permeable membranes, to supply nutrients and gases and to remove products and toxic metabolites. These systems are discussed further in Chapter 10.

Bioreactor design for plant-cell culture must take account of the unique characteristics of plant cells: large size, shear sensitivity, slow growth, product formation during slow growth or stationary phases, and the requirement for a certain amount of cell-cell contact or differentiation for favourable production of secondary metabolites. Suspension culture is used for industrial production of shikonin.

## Solid substrate fermentations

Solid substrate fermentations, used for cultivation of micro-organisms on solid materials containing no or a limited amount of free water, are used in oriental fermentations, fungal enzyme production by surface culture, mushroom production and in other applications. The most common substrates used are wheat bran, cereal grains, legume seeds, wood and straw. Micro-organisms which grow well in solid substrate fermentations are usually organisms that can tolerate a low water activity factor $(A_w)$.

Micro-organisms respond differently to water activity. By reducing $A_w$ below 0.95, bacterial growth is inhibited. Fungi and yeasts can grow at an $A_w$ of around 0.7. Solid substrate fermentations take different forms depending on whether the micro-organisms used are indigenous or pure isolated cultures. Composting for mushroom production involves the successive activities of a range of populations of indigenous micro-organisms from mesophilic bacteria, yeasts and moulds through to thermophilic fungi and actinomycetes. The traditional Koji process for fermentation of grains and soya beans and similar processes for production of mould industrial enzymes, both using cultures of *Aspergillus oryzae* and related species, are examples of solid substrate fermentations involving pure fungal cultures.

Since most industrial solid substrate fermentations are aerobic, fermentation conditions must be designed to promote efficient oxygen transfer and $CO_2$ removal from the substrate medium. Because of the high concentration of substrate per unit volume, heat generation during fermentation is usually much greater per unit volume than in liquid fermentations and the lower moisture content makes heat removal more difficult. Choice of substrate particle size is of critical importance in order to optimize inter-particle void spaces to facilitate gas and heat transfer. Heat removal can be facilitated by increasing the aeration rate of the system.

Advantages of solid substrate fermentation systems for industrial fermentations may include superior productivity, simpler technique, low capital investment, reduced energy requirement, low wastewater output, and lack of foam problems. Limitations often encountered include heat build-up, bacterial contamination, problems of scale-up and difficulty of controlling substrate moisture level.

Examples of designs of solid substrate fermenters include slow continuous agitation systems (such as rotating drums), tray systems and air-flow systems, where conditioned air is blown through the substrate bed in a cultivation chamber (Fig. 3.11). Rotating drums are usually equipped with an inlet and outlet for circulation of humidified air and often contain baffles or sections to agitate the contents. Tray fermenters holding 1–2 inch deep layers of substrate are stacked in chambers usually force aerated with humidified air. In the forced-air cultivation chamber, bed temperature is monitored and the appropriate temperature adjustment is made to the recycling air flow.

## Instrumentation and control

Instrumentation is used on fermenters to facilitate the analysis and recording of specific parameters, to aid in establishing optimum conditions for the fermentation and to maintain the optimized production process. Control of a particular parameter involves a sensor which can measure the property and a controller which compares the measurement with a predetermined set point and activates

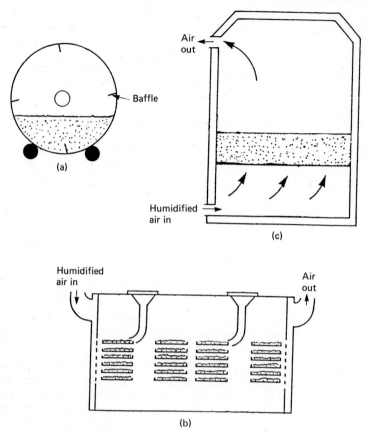

**Fig. 3.11**   Solid substrate fermenter designs. (a) Rotating drum, (b) tray system, (c) forced air-flow system.

equipment to adjust the property back towards the set-point. The adjustment usually involves modification of a valve opening or pump speed setting.

Sensors may be on-line, that is connected to the fermenter installation or in contact with the process stream, or off-line, where a sample is aseptically removed for analysis. Common fermenter on-line sensors are used for physical measurement of temperature, pressure, impeller r.p.m., liquid and gas flow rates and for physico-chemical measurement of pH and gas concentrations in the liquid and gas phases. On-line sensors which are in contact with the fermenter medium, including pH and dissolved gas probes, should be steam-sterilizable, be easy to calibrate and give a reliable continuous read-out. Some of the more important on-line measurement systems are described in Table 3.3.

Perhaps exhaust-gas analysis currently provides the greatest on-line insight into the state of the fermentation. Methods for $O_2$ analysis by paramagnetic procedures and $CO_2$ measurement by infrared analysis suffer from the disadvantage of calibration drift and slow response time. Mass spectrometers can

**Table 3.3** Fermenter on-line measurement systems

| *Parameter* | *Measuring equipment* | *Measurement principle (comment)* |
|---|---|---|
| Temperature | Electrical resistance thermometers and thermistors | Electrical resistance changes with temperature |
| Pressure | Bourdon tube pressure gauge | The elliptical cross-section of a partial hollow coil tends to become circular, gradually straightening the coil as the pressure increases inside |
| | Diaphragm pressure sensor | Movement of a diaphragm in response to pressure changes |
| | Electronic strain gauges | Electrical resistance changes when a wire is subject to strain |
| Vessel contents | Load cell measurement (weight) | Vessel weight is measured by a strain gauge, whereby electrical resistance is proportional to load |
| | Liquid level-sensing using capacitance probes (volume) | Probe capacitance changes with liquid level |
| Foam | Metal foam probe insulated at tip set at defined level above broth surface | Foam touching probe tip completes an electrical circuit actuating an antifoam feed device (moisture condensation along the outside of the insulation can short the system) |
| Impeller r.p.m. | Tachometer | Detection mechanism using induction, voltage generation, light sensing or magnetic force |
| Gas flow rate | Rotameter | The position of a free-moving float in a vertically-mounted tube with increasing bore is graduated to indicate flow rate |
| | Thermal mass flow meter | Detection of a temperature differential across a heater device placed in a gas flow path |
| Liquid flow | Propellers, turbines, rotameter | Systems graduated to indicate flow rate (not suitable for aseptic use) |
| | Electrical flow transducer | Liquid flow in a magnetic field such that the voltage induced is proportional to fluid relative velocity and the magnetic field. (May be used aseptically) |

**Table 3.3** continued

| Parameter | Measuring equipment | Measurement principle (comment) |
|---|---|---|
|  | Diaphragm-type metering pumps and peristaltic pumps | Indirect measurement of flow rate using pre-calibrated pumps. (Suitable for aseptic measurement and pumping) |
| pH | Combined glass reference electrode | Potentiometric measurement of hydrogen ion concentration |
| Dissolved $O_2$ | $O_2$ electrode | Electrons produced by interaction of oxygen with a metal surface generates a current. Probe-sensing tip contains a membrane through which oxygen can diffuse |
| Dissolved $CO_2$ | $CO_2$ electrode | Sensor is a pH probe surrounded by a $HCO_3^- + H^+$ solution, the pH of which is influenced by $HCO_3^-$ ions. Sensing tip contains a $HCO_3^-$ permeable membrane. (Electrodes expensive and unreliable) |
| Gas-phase $O_2$ | Paramagnetic gas analysis | Deflection and thermal analysis systems based on the strong affinity of oxygen for a magnetic field |
| Gas-phase $CO_2$ | Infrared analysis | $CO_2$ absorbs in the infrared range. (Measurement of gas-phase $O_2$ and $CO_2$ is usually carried out at the fermenter exhaust.) Normally it is assumed that the $O_2$ and $CO_2$ concentrations of incoming air are 20.91% and 0.03%, respectively |
| General gas analysis | Mass spectrometry | Gas molecules entering the instrument are ionized and accelerate through a magnetic field. Paths of gas molecules of different masses bend to different degrees and are collected and measured |

operate unattended for long periods and can measure $CO_2$, $O_2$, $N_2$ and other gases to a high degree of accuracy and when linked to a computer can generate valuable data such as respiratory quotients ($RQ = CO_2$ produced$/O_2$ consumed) and oxygen uptake rate. The fast response rate enables some mass spectrometers to

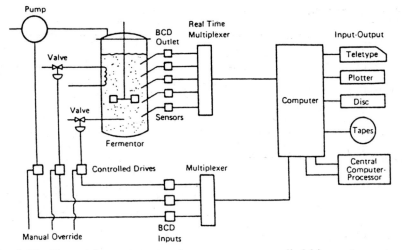

**Fig. 3.12** Essential components of a computer-controlled bioreactor system (reproduced with permission from Wilson, 1984).

be linked to many fermenters, via a sampling manifold, so that frequent data points can be obtained from each gas stream.

While off-line analysis is still commonplace for measurement of many substrates, metabolites, enzymes, cell constituents and biomass, the trend towards more automated process control creates an obvious demand for methods which can quantitatively measure biological molecules on-line. Some existing automatic on-line monitoring systems may be applied to fermenters if the problems of aseptic sampling from the liquid phase can be solved by use of microfiltration sampling techniques. HPLC on-line monitoring is possible using filtration or dialysis methods.

A new generation of highly specific biosensors has been developed, predominantly with clinical applications in mind, by interfacing immobilized enzymes with electrochemical sensors. A sterilizable glucose electrode and a sensitive alcohol electrode are examples of systems currently available for process control. The development of on-line biosensors which take advantage of the conformational changes occurring in enzyme proteins as they interact with substrates, or the electron donor/acceptor properties of biological molecules will undoubtedly lead to a greater level of sophistication in fermentation process control.

Computers may be used in fermentation processes to log data coming from sensors. They can analyse or process the data, present the analysis on display devices and store it or use it for process control by signalling activation switches, valves and pumps. Data processing operations would include calculation of rates, yields, productivity, respiratory quotients, etc. The storage and organization of fermentation data using computers is an extremely important component of most industrial fermentation processes. Apart from maintaining proper batch records for quality-assurance purposes and for inspection by regulatory agencies, significant process improvements are often achieved by judicious analysis of process records. Process control systems may simply involve automatic implementation of sequential operations such as a sterilization cycle. In addition, environmental properties may be maintained at set-points using digital set-point control whereby the computer compares the output and set-point controller. Alternatively environmental conditions may be maintained using direct digital control, whereby the sensors are interfaced directly with the computer and individual control units are not required. Key elements of a computer-controlled fermentation system are illustrated in Fig. 3.12.

A fully computerized integrated fermentation system requires detailed process models which can detect and respond to changes in culture conditions which may influence cell physiology and productivity. Computers may also be used to implement fermentation research and development strategies for optimization of growth and product formation. Finally, scheduling of the overall operation of a multi-unit fermentation plant, including upstream and downstream processing, can be mediated by a centralized computer.

# Chapter 4

# *Fermentation raw materials*

## Criteria used in media formulation

Media used in industrial fermentations must contain all elements required for synthesis of cell materials and for product formation. In addition media must be designed to satisfy the technical objectives of the process by providing a favourable environment for growth and/or product formation while at the same time being cost-effective. Many fermentation processes have several stages—development and scale-up of inocula, microbial growth in the production fermenter, product formation in the production fermenter—with each stage having separate technical objectives and medium requirements. In inoculum development and scale-up stages, the objective is usually to achieve high growth rates and to prepare high levels of viable biomass in a suitable physiological form for use as an inoculum for the next stage. Where viable biomass or single-cell protein (SCP) is the required end-product, the production fermenter medium and conditions are designed to achieve high growth rates and high biomass yields of viable cells or cells of the appropriate composition for SCP. Where the end-product is a cell product, such as an enzyme (or other protein) or metabolite, the situation is more complex. If the product is growth-associated, medium conditions have to simultaneously satisfy cell-growth and product-formation requirements in an optimized manner. Where the bulk of product formation occurs after cell growth, medium conditions may be designed such that initially cells having the appropriate physiological properties are efficiently produced during the exponential phase after which conditions must favour synthesis of product at optimal rates and maximum final product yields. With many industrial products produced predominantly after the exponential phase of growth, for example enzymes and antibiotics, a linear optimum rate of product formation is often observed during the post-exponential phase and a key objective is to control media and other environmental conditions such that this linear production phase is extended for as long as possible.


## INFLUENCE OF MEDIUM ON CELL GROWTH

Micro-organisms require carbon, nitrogen, minerals, sometimes growth factors, water and (if aerobic) oxygen as elements for cell biomass and as energy for biosynthesis and cell maintenance. The elemental composition of most micro-organisms is fairly similar and consequently can be used as a starting point to designing an optimally-balanced medium for fermentation. Typical values for C, H, O, N and S as a percentage of dry cell weight are 45, 7, 33, 10 and 2.5, respectively. Trace elements such as Cu, Mn, Co, Mo, B and possibly other metals may also be required, depending on the water source. Some micro-organisms grow on media which contain no growth factors while others require complex media containing specific nutrients such as amino acids, vitamins or nucleotides. Organisms which do not require these supplements often have substantially higher growth rates in complex media. Specific growth-factor supplements may be added to media as pure chemicals, for example use of biotin in glutamic acid fermentations, or are more usually present as components of cruder nitrogen sources such as corn steep liquor. In practically all industrial fermentations, the carbon substrate provides energy for growth as well as carbon for biosynthesis. The carbon requirement under aerobic conditions may be determined from the biomass yield co-efficient

$$Y_c = \frac{\text{Biomass produced (g)}}{\text{Carbon substrate utilized (g)}}$$

The effects of media-related environmental conditions such as pH, substrate concentration, etc., on growth rate have already been discussed (Chapter 2).


## PRODUCT FORMATION

Where product formation is dependent on inducers, the specific inducer or a structural analogue must be incorporated into the medium and perhaps maintained there by continuous or pulsed additions during the process. Likewise, if product formation is subject to catabolite repression by substrates such as glucose and other carbohydrates, it is necessary either to use a de-regulated mutant or to supply the carbohydrate by continuous or pulsed feeding, such that its level remains below the critical concentration required for repression. Many enzyme production processes are subject to induction and/or catabolite repression. Low productivity in secondary metabolite fermentations is often associated with the presence in the medium of high concentrations of rapidly metabolizable carbon sources.

A rapidly utilized nitrogen source inhibits production of some antibiotics. In fungi, the ammonium ion represses uptake of amino acids by general and specific amino acid permeases. Ammonia also represses nitrate assimilation in many micro-organisms, so that nitrate is only used as an alternative nitrogen source when the ammonia has been depleted.

The concentrations of certain minerals are critical in some fermentations. Many secondary metabolite fermentations require the concentration of inorganic phosphate to be low during the production phase and calcium sometimes plays a role in precipitating excess phosphate. The concentrations of trace elements such as iron and/or zinc have a critical effect on the production of penicillin, actinomycin, chloramphenicol, neomycin, griseofulvin and riboflavin. Manganese concentration is important in bacitracin and citric acid production. Some of the effects of these metals have been explained in terms of activation or inhibition of enzyme activity.

In some fermentations, for example in the production of some antibiotics, vitamins and amino acids, precursors are added to the medium which are directly incorporated into the product. Phenylacetic acid is a major precursor used in the production of benzylpenicillin by fermentation.

The optimized production of some metabolites requires the incorporation of specific inhibitors in the medium either to minimize formation of other metabolic intermediates or to prevent further metabolism of the desired product. In tetracycline production by *Streptomyces* species, bromide represses chlorotetracycline production, and in glycerol production by *Saccharomyces cerevisiae*, sodium bisulphite suppresses reduction of acetaldehyde, so that the re-oxidation of $NADH + H^+$ involves the conversion of dihydroxyacetone phosphate to glycerol-3-phosphate, which in turn is converted to glycerol.

OXYGEN

Most fermentation processes are aerobic requiring provision of oxygen as a raw material. Complete oxidation of glucose represented by the equation

$$C_6H_{12}O_6 + 6O_2 = 6H_2O + 6CO_2$$

requires 192 g oxygen to oxidize 180 g glucose. Assuming that no extracellular product other than $CO_2$ and $H_2O$ is formed, the oxygen requirement to produce 1 g of bacterial dry weight from glucose is approximately 0.4 g.

In addition to this overall demand for oxygen, metabolic processes are affected by medium-dissolved oxygen concentration. The relationship between dissolved oxygen concentration and specific oxygen uptake rate ($Q_{O_2}$, mmoles $O_2$ consumed per gram dry weight of cells per hour) follows the typical Michaelis–Menten pattern (Fig. 4.1). Thus in order to maintain maximum biomass production, the dissolved oxygen level must be kept above the critical concentration ($C_{crit}$). Dissolved oxygen levels for optimum product formation rate may be quite different from the $C_{crit}$ concentrations. In general, dissolved-oxygen concentrations in excess of $C_{crit}$ appear to be required to support metabolite production arising from tricarboxylic acid intermediates, whereas oxygen concentrations

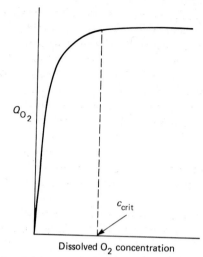

**Fig. 4.1** Effect of oxygen concentration on specific oxygen-uptake rate $(Q_{O_2})$ of a micro-organism.

below $C_{crit}$ tend to support non-tricarboxylic acid metabolite production from phosphoenolpyruvate and pyruvate.

COMBINED EFFECTS

The effects of medium constituents and oxygen on growth and product formation are often inter-related and complex. The observation that the yield of yeast biomass from sugar is greater and the yield of alcohol is reduced in the presence of oxygen as compared with anaerobic conditions is known as the Pasteur effect. However, at concentrations of glucose greater than 250–300 mg l$^{-1}$, even under aerobic conditions, the oxidative enzymes of the tricarboxylic acid cycle and of the cytochrome systems are catabolite repressed and biomass yields remain low with most of the sugar being converted to alcohol. In this case a reduced yield of biomass results from catabolite repression. In industrial fermentations for production of primary and secondary metabolites, over-production of biomass reduces product yields. With primary metabolites, too great a proportion of substrate being converted to biomass reduces metabolite product yields. Also, production of some metabolites appears to require a stunted cell physiology. Over-production of biomass delays the onset of secondary metabolite production. High fungal biomass concentrations can be deleterious to oxygen-required product-formation processes by limiting oxygen-transfer rates.

PHYSIOLOGY AND MORPHOLOGY

Medium constituents can have a profound effect on the physiology of micro-organisms and optimized product formation is often associated with particular physiological forms. In optimized fermentations for citric acid production by

*Aspergillus niger*, the mycelia occur as small hard mycelial pellets and hyphae appear abnormally short and stubby. For itaconic acid production by *A. terreus*, pelleted growth is also preferred. Higher yields of fumaric acid were obtained from *Rhizopus arrhizus* when it was grown in a filamentous form. Filamentous growth is generally also preferred for penicillin production by *Penicillium chrysogenum*. Rates of mycelial growth and nutrient assimilation are generally significantly higher with filamentous growth rather than pelleted growth. Consequently, medium conditions which induce filamentous growth are preferred for biomass production.

With streptomycin-producing strains of *Streptomyces griseus*, mycelial fragmentation and loss of conidia formation, and a progressive change to nocardial-type morphologies, is associated with culture degeneration and antibiotic-producing capacity.

Medium-related conditions which have been shown to affect morphology include pH, viscosity, divalent cations, chelating agents, anionic polymers, surface-active agents and the presence of solids in the medium. Hyphae of *P. chrysogenum* become shorter and thicker at more alkaline pH and subsequently form pellets. Polymeric substances added to the medium tend to cause dispersed filamentous growth. Carbopol, a carboxypolymethylene polymer, alginic acid and carboxymethylcellulose have been shown to induce filamentous growth in species such as *P. chrysogenum*, *A. niger* and *R. arrhizus*. Divalent cations often cause pelleted growth and their effect can be counteracted by simultaneous addition of chelating agents or ion-exchange materials such as anionic polymers. Many observations suggest that divalent cations may contribute to pellet formation by interacting with anionic groups on cell surfaces, thereby linking adjacent cell walls by salt-bridge formation or by overcoming the repulsive forces of negatively-charged surfaces to facilitate pelleting. Cell–cell interactions leading to pellet formation sometimes occur between spores prior to germination whereas in other cases the presence of mycelial hyphae is required. Clearly these charge-dependent interactions can be influenced directly by medium ionic constituents. Environmental conditions, including media, which alter the composition of the cell surface may also be expected to influence morphology. Cell-wall composition of *A. niger* has been shown to be significantly influenced by manganese (Table 4.1).

**Table 4.1** Effect of manganese on cell-wall composition of *A. niger* (calculations from data of Kisser *et al.*, 1980)

| Cell-wall composition (%) | Manganese deficient | Manganese supplemented |
|---|---|---|
| α-Glucan | 66.4 | 59.0 |
| β-Glucan | 7.0 | 18.2 |
| Chitin | 20.0 | 6.8 |
| Protein | 2.4 | 2.9 |
| Galactomannan ⎫<br>Galactosamine ⎭ | 3.7 | 12.6 |
| Lipids | 0.5 | 0.5 |

The presence of cations has been shown to affect yeast flocculation and yeast cellular components have been implicated in cross-linking of yeast cells. The phenomenon of yeast flocculation is important in the sedimentation and removal of yeast in production of non-distilled alcoholic beverages. Use of flocculent rapid-settling strains of yeast in continuous fermentation processes for alcohol production, using upward flowing vertical fermenters, enables the yeast to settle back with separation from the rising beer, thereby increasing the cell density in the fermenter without use of centrifuges.

Production of some microbial products is associated with spore formation while synthesis of other products is inhibited by spore formation. Sporulation can be regulated by media design. In general, low levels of complex nitrogen induce spore formation whereas high amino acid concentrations have an inhibitory effect.

Culture media may be manipulated to affect cell permeability. Biotin limitation in glutamic acid fermentations inhibits complete biosynthesis of oleic acid by glutamic acid producing bacteria and modifies cell permeability so that glutamic acid is excreted into the medium. Antibiotics, including penicillin, which inhibit the formation of peptidoglycan cross-links, result in formation of swollen cells which also leak glutamic acid into the medium. Some surface-active agents have been shown to increase the rate of secretion of microbial extracellular enzymes.

## RAW MATERIALS FOR PROCESS CONTROL

Fermentation raw materials used directly for process control include buffers, acids and alkalis for pH control, antifoams to suppress foam and nutrient feeds to circumvent phenomena such as catabolite repression.

### *pH control*

Inorganic phosphate is often added to fermentations at levels in excess of those required for growth as it has good buffering capacity in the region pH 6.0–7.5. Organic acids can be used to provide buffering capacity at lower pH values. Media may also be very efficiently buffered against acid production by use of calcium carbonate. The pH may be controlled also by addition of hydroxide salts, liquid or gaseous ammonia, and sulphuric and hydrochloric acids.

Medium pH is often controlled indirectly by judicious balancing of carbohydrate and nitrogen sources. Carbohydrates often contribute to reducing pH due to the formation of organic acids. Ammonium salts usually produce acid conditions due to liberation of the free acid during assimilation of the ammonium ion. Nitrate assimilation will cause an alkaline drift but when added as ammonium nitrate, the ammonium ion is preferentially assimilated. Organic sources of nitrogen, such as hydrolysed proteins, corn steep liquor and distiller's solubles may cause the pH to drift towards the alkaline side. In manipulating medium pH in this manner, account has to be taken of the possible repressive effects of the constituents being used on product formation.

*Foam*

Foam frequently occurs in fermentations due to the denaturation of proteins at the gas–liquid interface. If uncontrolled, the foam may rise to fill the fermenter headspace and can eventually result in the evacuation of much of the fermenter contents through the air exhaust outlet. Foaming can also cause removal of cells from the medium.

Antifoams are surface-active agents which act by reducing the surface tension in foams to disperse the foam. Antifoams vary in their effectiveness with fermentation conditions (medium composition, microbial strain, stage of growth, aeration pattern and fermenter configuration) and selection of antifoam type is often based on trial-and-error procedures. Many antifoams have low solubility and require a carrier such as a natural or mineral oil.

Use level of antifoam should be minimized as antifoams can affect oxygen-transfer rates by as much as 50%. In addition, the antifoam should be non-toxic, non-hazardous, heat-sterilizable and cheap. Mechanical foam breakers may be used as an alternative to antifoams.


## IMPLICATIONS FOR DOWN-STREAM PROCESSING

Design of fermentation media have knock-on implications for down-stream processing. Use of cruder fermentation materials may save in nutrient costs but results in a more complex and costly recovery process. High usage rates of antifoam can result in recovery of antifoam with solids during the first solid–liquid separation stage and can be problematic in the processing of cell-associated products. Selection of fermentation raw materials must therefore take into account the economic impact on fermentation productivity, product purification and waste treatment. In general, in processes which are fermentation cost-intensive, medium cost will represent a substantial proportion of total manufacturing costs and there is a particular need to use inexpensive raw materials. Fermentation processes that are recovery cost-intensive tend to use purer raw materials in order to minimize recovery problems.


## Survey of microbial fermentation media

Petrochemicals, such as hydrocarbons, alcohols and acids, once provided feedstocks for fermentation when crude oil was cheap but are no longer used to a significant extent. Currently, the most significant fermentation carbon and energy sources are renewable raw materials containing sugar and starch and to a lesser extent fats and oils. Competition exists between use of starch and starch-based materials for food production and as a feedstock, since the current annual deficit of foodstuffs in developing countries amounts to about $10^8$ t. Lignocellulose constitutes about 50% of a world annual yield of biomass of approximately $10^{11}$ t. In contrast, annual starch and sugar yields are far smaller, amounting to approximately $10^9$ and $10^8$ t, respectively. In the medium to long-term, therefore, hydrolysis products of lignocellulose must become important fermentation feedstocks.


CARBOHYDRATES

Starch is the most important carbohydrate currently used in fermentation processes. It may be used in the form of whole or crushed cereals or roots from plants such as corn, rice, wheat, potatoes and cassava, as purified starch or as modified starches or dextrins. During heating or sterilization of starches, the starch gelatinizes and becomes extremely viscous and usually an enzymic starch hydrolysis step is included in the process to liquefy or thin the starch. This involves use of amylases from microbial sources or malted cereals. The extent of starch hydrolysis required varies with fermentation process and depends on considerations as to whether or not the microbial strain to be used produces amylases and whether product synthesis is subject to catabolite repression.

Cellulose present in wood is usually combined with hemicellulose and lignin in the form of lignocellulose. The lignin makes the cellulose resistant to microbial attack, and so far, chemical and enzymatic methods for conversion of lignocellulose into fermentable sugars are not cost-effective. There is, however, limited use of lignocellulose in mushroom fermentations and refined cellulose is used as a substrate for production of cellulolytic enzymes.

Sucrose is available for use in fermentation processes either in crystalline form or in crude form in raw juice or in molasses, a by-product of sugar manufacture. While sugar contained in molasses is obviously cheaper, the composition of molasses varies notoriously with source (cane or beet), quality of the crop and the nature of the sugar-refining process, causing problems of fermentation reproducibility.

Lactose is present in whey at a concentration of 4–5% and whole whey or deproteinized whey is used as a cheap source of carbohydrate in some alcohol-production processes. The low carbohydrate concentration of whey makes transporting of the material prohibitively expensive and consequently whey fermentations are usually carried out near the source of whey, the cheese factory.

Glucose is usually produced in fermentation media by direct enzymic conversion of starch. In some instances more expensive refined glucose, as syrup or in crystalline form, is used for production of higher value products.

Plant oils such as soybean oil, palm oil and cotton seed oil may be used in fermentation media to supplement carbohydrates. Methanol, one of the cheapest fermentation substrates, is used for SCP production, but as it is used only by a few bacterial and yeast strains, its wider application in fermentation processes has been limited. Ethanol, which can be metabolized by many micro-organisms as sole carbon source, may be a potential starting material for production of other fermentation products in the future. Acetic acid is currently produced by microbial oxidation of ethanol.

**Table 4.2** Complex substrates frequently used in microbial fermentation media (adapted from Miller and Churchill, 1986)

| Source | Ingredient | Dry matter | Protein | Carbohydrates | Fat | Fibre | Ash |
|---|---|---|---|---|---|---|---|
| | | | | *Major components*(%) | | | |
| Whole cereals | Barley | 90.0 | 11.5 | 68.0 | 1.8 | 7.0 | 2.5 |
| | Barley malt | 96.0 | 13.0 | 70.0 | 2.0 | 3.5 | 2.5 |
| | Corn | 82.0 | 9.9 | 69.2 | 4.4 | 2.3 | 1.3 |
| | Oats | 86.5 | 12.0 | 54.0 | 4.5 | 12.0 | 4.0 |
| | Rice | 89.5 | 8.0 | 65.0 | 2.0 | 10.0 | 4.5 |
| | Wheat | 90.0 | 13.2 | 69.0 | 1.9 | 2.6 | 1.8 |
| Plant by-products | Beet molasses | 77.0 | 6.7 | 65.1 | 0.0 | 0.0 | 5.2 |
| | Beet pulp | 90.0 | 8.9 | 59.1 | 0.6 | 18.3 | 3.1 |
| | Blackstrap molasses | 78.0 | 3.0 | 54.0 | 0.4 | – | 9.0 |
| | Citrus pulp (dried) | 90.0 | 6.0 | 62.7 | 3.4 | 13.0 | 6.9 |
| | Corn germ meal | 93.0 | 22.6 | 53.2 | 1.9 | 9.5 | 3.3 |
| | Corn gluten meal (60%) | 90.0 | 62.0 | 20.0 | 2.5 | 1.6 | 1.8 |
| | Corn steep liquor | 50.0 | 24.0 | 5.8 | 1.0 | 1.0 | 8.8 |
| | Corn steep powder | 95.0 | 48.0 | – | 0.4 | – | 17.0 |
| | Cotton seed meal | 94.0 | 41.0 | 28.9 | 3.9 | 13.5 | 6.7 |
| | Dried distillers solubles | 92.0 | 26.0 | 45.0 | 9.0 | 4.0 | 8.0 |
| | Linseed meal | 92.0 | 36.0 | 38.0 | 0.5 | 9.5 | 6.5 |
| | Rice bran | 91.0 | 13.0 | 45.0 | 13.0 | 14.0 | 16.0 |
| | Soybean meal | 90.0 | 42.0 | 29.9 | 4.0 | 6.0 | 6.5 |
| | Soybean meal (low fat) | 90.0 | 45.0 | 32.2 | 0.8 | 6.5 | 5.5 |
| Animal by-products | Blood meal | 93.0 | 80.0 | 2.5 | <1.0 | <1.0 | 3.0 |
| | Fish meal (herring) (70%) | 93.0 | 72.0 | – | 7.5 | 1.0 | – |
| | Meat and bone meal | 92.0 | 50.0 | 0.0 | 8.0 | 3.3 | 31.0 |
| | Whey (dried) | 95.0 | 12.0 | 68.0 | 1.0 | 0.0 | 9.6 |
| Microbial by-products | Yeast hydrolysate | 94.6 | 52.5 | – | 0.0 | 1.5 | 10.0 |

**Table 4.3**   Functions attributed to serum in cell-culture media

- Supplies hormones and growth factors necessary for cell function
- Provides factors necessary for substrate attachment
- Acts as a pH buffer
- Binds and inactivates or sequesters toxic compounds
- Contains binding proteins which stabilize and/or deliver nutrients and hormones to the cell
- Supplies nutrients for cell metabolism
- Contains protease inhibitors
- Contains trace elements (i.e. selenium)

NITROGEN

The most important sources of nitrogen for fermentation are ammonia, nitrates, urea and nitrogen present in complex cereals, root crops and their by-products. Purified amino acids are only used in very special cases in fermentation processes, usually as precursors. Complex substrates also often supply vitamins, growth factors and minerals which have a key influence on the fermentation process.

COMPLEX SUBSTRATES

Crude complex substrates, therefore, provide a cheap source of carbon, nitrogen and other nutrients for fermentation. They include whole plant tissues and an array of plant, animal and microbial by-products. Some of the major complex substrates used in fermentation processes are listed in Table 4.2.

## Animal cell-culture media

Historically, foetal calf serum has been incorporated into basal media for mammalian cell culture use as a source of essential growth factors. However, its supply is limited and price is high. Other sera which are widely used include calf serum, newborn calf serum, and horse serum. While foetal calf serum is most universally applicable, the potential volume of these other sera is much greater. Substantial savings have been achieved by blending foetal serum with alternative sera or by supplementing reduced serum levels with media having fortified growth factors. Normally, serum constitutes 5–10% by volume of culture media, but if the basal medium is optimized for a particular cell type, this can be reduced to 1–2%. Serum contains several factors essential for cellular competence and proliferation, including initiation factors, binding proteins, attachment factors and low-molecular-weight compounds. The main functions of serum in cell-culture media are listed in Table 4.3.

The composition of a formulated basal medium, BME (basal medium Eagle's),

**Table 4.4** Composition of BME (basal medium Eagle's) cell-culture medium

| Inorganic salts | Concentration (mg/l) | Amino acids | Concentration (mg/l) | Vitamins | Concentration (mg/l) |
|---|---|---|---|---|---|
| $CaCl_2$ | 200.0 | L-arginine | 17.4 | Biotin | 1.0 |
| KCl | 400.0 | L-cystine | 12.0 | D-Ca pantothenate | 1.0 |
| $MgSO_4 \cdot 7H_2O$ | 200.0 | L-glutamine | 292.0 | Choline chloride | 1.0 |
| NaCl | 6800.0 | L-histidine | 8.0 | Folic acid | 1.0 |
| $NaHCO_3$ | 2200.0 | L-isoleucine | 26.0 | i-Inositol | 1.0 |
| $NaH_2PO_4 \cdot H_2O$ | 140.0 | L-leucine | 26.0 | Nicotinamide | 1.0 |
| | | L-lysine | 29.2 | Pyridoxal·HCl | 1.0 |
| | | L-methionine | 7.5 | Riboflavin | 1.0 |
| | | L-phenylalanine | 16.5 | Thiamine·HCl | 1.0 |
| | | L-threonine | 24.0 | | |
| | | L-tryptophan | 4.0 | *Other compounds* | |
| | | L-tyrosine | 18.0 | D-glucose | 1000.0 |
| | | L-valine | 23.5 | Phenol red | 10.0 |

**Table 4.5** Additional components which may be present in more enriched media

| Inorganic salts | Amino acids | Vitamins | Other compounds |
|---|---|---|---|
| $CuSO_4 \cdot 5H_2O$ | L-Alanine | L-Ascorbic acid | Sodium pyruvate |
| $FeSO_4 \cdot 7H_2O$ | L-Asparagine | Niacinamide | Glutathione |
| $ZnSO_4 \cdot 7H_2O$ | L-Aspartic acid | p-Aminobenzoic acid | Lipoic acid |
| $Fe(NO_3)_3 \cdot 9H_2O$ | L-Cysteine·HCl·$H_2O$ | Vitamin $B_{12}$ | Linoleic acid |
| $KNO_3$ | L-Glutamic acid | | Hypoxanthine |
| $Na_2SeO_3 \cdot 5H_2O$ | Glycine | | Thymidine |
| | L-Proline | | Putrescine·2HCl |
| | L-Hydroxyproline | | Hydroxyethylpiperazine |
| | L-Serine | | |

**Table 4.6**   Selected growth factors for possible use in serum-free media

| | |
|---|---|
| ● Colony-stimulating factor | ● Platelet-derived growth factor |
| ● Estradiol | ● Prolactin |
| ● Epidermal growth factor | ● Prostaglandins |
| ● Fibroblast growth factor | ● Retinoic acid |
| ● Fibronectin | ● Selenium |
| ● Growth hormone | ● Serum albumin |
| ● Hemin | ● Somatomedin |
| ● Hydrocortisone | ● T-cell growth factor |
| ● Insulin | ● Transforming growth factor |
| ● Nerve cell-growth factor | ● Transferrin |
| ● Phospholipids | ● Triiodothyronine |

widely used in mammalian cell culture is given in Table 4.4. Formulated basal media such as BME have relatively high concentrations of the more common nutrients and are ideal for encouraging proliferation. More complex media, for example Ham's F10./F12. have a greater range of nutrients but concentrations of some constituents are reduced somewhat to maintain osmotic balance. Practically all of the ingredients present in BME are also contained in more enriched media formulations. Some of the supplementary components present in selected more enriched media are indicated in Table 4.5. Primary mammalian cells, derived from human, bovine, equine and other primate sources appear to proliferate best in basal culture media and growth may be inhibited by use of more highly-enriched media. Media of intermediate nutrient supplementation are generally used for growth of secondary mammalian cells and immortal cell lines. More highly-enriched media are usually used for rapidly-proliferating cells. More complex media are sometimes needed in order to allow full expression of a differentiated function. Differentiation usually implies cessation of proliferation.

Use of serum free or chemically-defined media allows researchers to investigate cell-culture processes with minimal extraneous interference and facilitates isolation and purification of mammalian cell-culture products. Serum free media must contain the components of serum necessary for cell culture and serve the main functions listed in Table 4.3. By adapting cells from minimal growth media to more complex formulations the nutritional contribution of serum may be substituted. The growth factor combinations required for mammalian cell culture tend to be cell-specific. A general list of growth factors which might be tested in the development of serum free media is given in Table 4.6.

**Nutrient media for growth of plant cells**

The majority of plant-cell culture media have a chemically-defined composition. Media contain an organic carbon source, a nitrogen source, inorganic salts and growth regulators. Sucrose is most frequently used as a carbon source but other mono- and di-saccharides including glucose, fructose, maltose and lactose have also been used. Nitrate is considered the most important nitrogen source in the medium although supplementation with ammonium salts is often beneficial. Some cell species require organic nitrogen in the form of amino acids. The inclusion of organic nitrogen often makes a positive impact in the early stages of culture initiation. Plant hormones are required for most types of plant-cell culture. Auxins induce cells to divide. Cytokinins, which have growth-regulatory roles in plants, are often used in combination with auxins in cell culture to promote cell division.

## Culture maintenance media

Media for storage and sub-culturing of key industrial strains should be designed to have the characteristics necessary to maintain good culture viability and to minimize genetic variation. Above all, it is essential that such media preserve the particular production capability for which the strain is used in industrial fermentations. In general, maintenance media should minimize production of toxic metabolites by the organism which might have strain-destabilizing effects. Strains which tend to be unstable with respect to their required production properties should if possible be maintained on media which are selective for that particular capability. Many of the cell-culture collections recommend mainten-ance media for each of their strains and these serve as a good starting point for maintenance media development. The catalogues of the *American Type Culture Collection* are an excellent source of formulations for culture maintenance media.

# Chapter 5

# *Down-stream processing*

## Introduction

This chapter on down-stream processing is concerned with the recovery and purification of the required product from the fermentation process. The objective in industrial fermentation processes is to recover and refine the product to the required specification, while achieving maximum product yield at minimum recovery costs. The nature of the recovery process will be influenced by the type of fermentation, physical and chemical properties of the product and unwanted by-products or contaminants, product concentration, product location (intra-cellular, extracellular), scale of operation, waste treatment implication, product stability and desired specification.

Fermentation and recovery operations are integral parts of an overall process. The nature of the fermentation process can significantly influence the recovery process. For example, fermentation cost savings or yield improvements achieved by use of more complex medium raw materials must take account of the possible implications for down-stream processing. For products which are difficult to separate from contaminating cell or medium materials, specific activities (activity of product per unit weight of biological material) may be more important than absolute product yields. Attempts should be made to eliminate or control production of undesirable by-products, which are difficult to separate from the principal fermentation product, by use of strain selection or modification techniques or by manipulation of fermentation conditions. Fermentation run-to-run variations with consequential production of variable broth or cell batches can create problems in down-stream processing. Contaminated fermentations are particularly problematic in this regard. Harvesting time is often critical to the stability of cells or product.

The concentration of the product in the starting material is a major factor in the

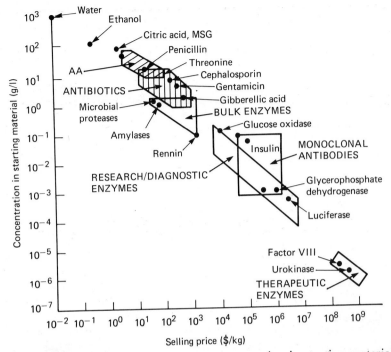

**Fig. 5.1** Relationship between product concentration in starting material and its selling price (reproduced with permission from Dwyer, 1984).

overall cost of production. The relationship between product concentration and selling price for a broad range of products illustrates this (Fig. 5.1). The influence of step yield and the number of steps on overall yield is illustrated in Fig. 5.2. Apart

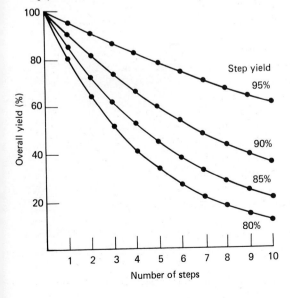

**Fig. 5.2** Effect of the number of purification steps and step yield on overall process yield (reproduced with permission from Fish and Lilly, 1984).

from the significant operating cost associated with each purification step, the cumulative yield loss observed in multi-stage purifications can be substantial even when average step yields are 80–90%. Hence for optimum economic viability, the number of purification steps required to attain the desired product specification should be minimized.

In addition, some biochemical purification processes require substantial analytical back-up support and some biological materials require complicated and time-consuming assay procedures. Such analytically-intensive purification steps should be avoided if possible as the analytical cost component may result in the purification step being cost-ineffective. In addition, time-lags due to monitoring extend overall processing time.

### Separation processes

The products of fermentation processes include gases, extracellular soluble molecules secreted into the fermentation broth and solids including cells containing intracellular soluble or insoluble molecules. Gases such as carbon dioxide and methane, produced for example during alcohol production and anaerobic digestion, may be collected from exhaust lines, purified and compressed for commercial use. In a limited number of cases, the whole fermentation broth is used directly or after evaporation or drying as the final product as in the case of some food fermentations and some surface culture enzyme fermentations. In other cases the product may be recovered from the whole fermentation broth by processes such as distillation, extraction, absorption and membrane separation. However, in the vast majority of cases, the first purification stage, for products which are either extracellular and soluble or intracellular, involves the separation of cells and solids from extracellular culture fluid. In the case of submerged culture fermentations this separation stage involves processes such as sedimentation, flocculation, centrifugation or filtration. With solid substrate fermentations and sometimes with highly viscous submerged culture broths, an aqueous extraction procedure precedes solid–liquid separation.

Following solid–liquid separation of the fermentation whole broth, each of the two fractions may be further processed or run to waste as appropriate. The solids, including cells, may be pressed into blocks or cakes, air or freeze-dried, extracted with solvents or disintegrated to release intracellular components by autolysis, chemical or mechanical methods. The disintegrated cells can be further processed into a solid and soluble fraction using solid–liquid separation techniques.

Many biotechnology products are soluble and are present either in the clarified extracellular culture fluid or in the intracellular soluble fraction. These products may be further processed and purified using molecular separation techniques which exploit property differences such as size, charge, solubility, volatility, biological affinity, etc.

Important distillation, evaporation and drying processes are summarized in Table 5.1. Distillation processes are used to recover volatile fermentation products such as ethanol and for recovery and re-use of solvents used in other down-stream

**Table 5.1** Examples of evaporation, distillation and drying processes

| *Process* | *Nature of equipment* | *Operating principle* |
|---|---|---|
| Evaporation | Long-tube vertical type evaporator and falling-film evaporators | Liquid evaporates as it passes through vertical tubes in contact with steam |
| Distillation | Batch and continuous distillation plants | Evaporation of solvent by heating the solution, separation of vaporized components based on volatility, recovery of volatile product by condensation |
| Drying | Spray drier | Liquid or paste is atomized into small droplets and passed through a hot gas stream resulting in rapid evaporation |
| | Drum drier | Solution is fed onto the surface of a heated, slowly rotating drum causing evaporation. Dried solids are discharged by use of a scraper blade |
| | Lyophilizer | Water is removed from frozen material by sublimation |

processing steps. Evaporation processes are used to produce liquid concentrates as final products or in the preparation of material for spray or drum drying. Conditions must take account of the heat sensitivity of the product. Spray-drying involves rapid evaporation and the evaporative cooling effect, at least in the initial stages of drying, protects against over-heating. It is a milder heat treatment process than drum-drying. Lyophilization is used for drying of heat-sensitive materials including live microbial cells and labile proteins.

Unit processes for separation of cells and solids from the liquid phase are summarized in Table 5.2. In unmixed containers, solids or cells may settle out naturally without chemical addition. Selected strains of brewing yeasts settle to the fermenter bottom at the end of fermentation. Flocculating agents which cause microbial cells or particles to aggregate, increase the sedimentation rate as larger particles tend to sediment at a faster rate. Efficient cell sedimentation is particularly important in some cell recycle fermentation processes where the sedimented cells are re-used. In many cases, however, flocculation–sedimentation processes are too time-consuming and would slow down the recovery of unstable products and extend fermentation turn-around time. In such cases centrifugation or filtration techniques are usually used.

**Table 5.2**  Fermenter cells/solid–liquid separation processes

| Process | Nature of equipment/example additives | Operating principle |
|---|---|---|
| Settling | Settling tank | Cells settle to vessel bottom by force of gravity |
| Flocculation | Isinglass, gelatin, tannic acid, divalent ions, quaternary ammonium compounds, synthetic flocculants | Cell or particle aggregation by neutralization of like or repulsive ionic surface charges or formation of bridges between particles by multivalent flocculants |
| Continuous centrifugation | Solids bowl separator | Solids retention in bowl. Feed solids 0–1% vol. |
| | Solids ejecting separator | Intermittent discharge. Feed solids 0.01–20% vol. |
| | Nozzle discharge separator | Continuous discharge. Feed solids 1–30% vol. |
| | Decanter-screw conveyor | Continuous discharge. Feed solids 5–80% vol. |
| Batch filtration | Plate and frame filter (pressure filter) | Stack of filter plates, cloth or pad covered, arranged such that slurry and filtrate flow to and from each plate respectively |
| Continuous filtration | Rotary vacuum filter | Slurry is fed to the outside of a drum filter. Filtrate is drawn through the filter into drum by vacuum. Filter cake is automatically removed using a string or scraper |
| | Cross-flow filtration | See Fig. 5.7 |

Choice of centrifugation equipment depends on the organism size and the required throughput. Separation efficiency has in many cases to be balanced against throughput. Centrifuges used for large-scale fermentation solids recovery have to be continuous and have a solids discharge mechanism. Nozzle discharge types are suitable for recovery of yeasts and bacteria but tend to get clogged by fungal mycelium or large particulate matter. Solids-ejecting centrifuges (Fig. 5.3), having continuous or intermittent mechanisms to discharge solids, may be used for recovery of mycelial or bacterial biomass. The screw-decanter centrifuge (Fig. 5.4), which is suitable for dewatering of coarse solid materials at high solids concentrations, has been used for recovery of yeasts and fungi.

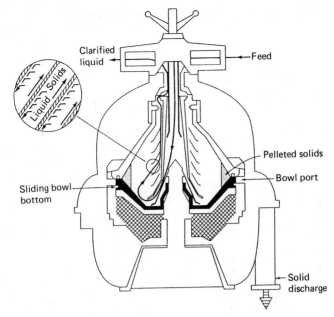

**Fig. 5.3** Solids discharge centrifuge (reproduced with permission from Bailey and Ollis, 1986).

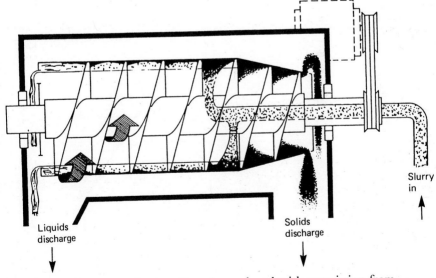

**Fig. 5.4** Screw-decanter centrifuge (reproduced with permission from Penwalt Corporation, Sharples-Stokes Division, Pa.

Plate and frame filters (Fig. 5.5) are cheap and versatile in that the surface area can be adjusted by varying the number of plates. They are not suitable for removing large quantities of solids from broths as the plates have to be dismantled for solids recovery. They are widely used as polishing devices to filter out low residual solids from broths or other liquids. Rotary vacuum filters (Fig. 5.6) are widely used to clarify large volumes of liquid with automatic solids discharge. Build-up of fine or gelatinous suspensions of bacteria or other materials on filters can slow down or block filtration. It is common practice to use filter aids to improve filter-cake porosity. The filter aid may be applied to the filter as a pre-

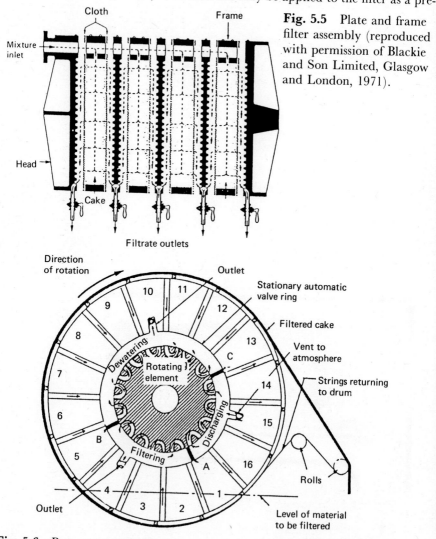

**Fig. 5.5** Plate and frame filter assembly (reproduced with permission of Blackie and Son Limited, Glasgow and London, 1971).

**Fig. 5.6** Rotary vacuum filter. Sections 1–4 are filtering; sections 5–12 are dewatering; the filter cake is discharged at section 13 (reproduced with permission from Ametek, Inc.).

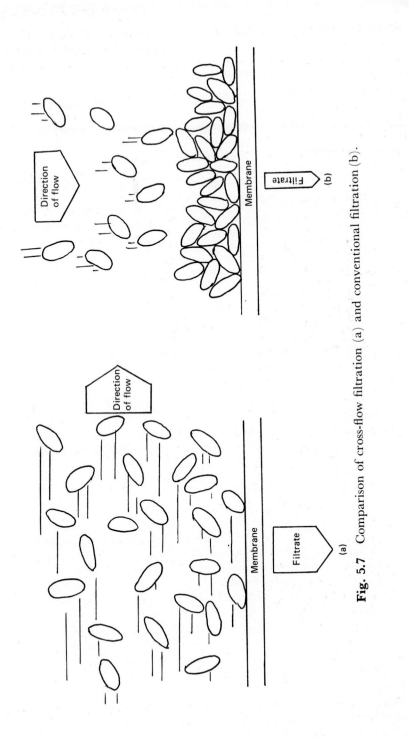

**Fig. 5.7** Comparison of cross-flow filtration (a) and conventional filtration (b).

**Table 5.3**  Important cell-breakage processes

| Process | Nature of equipment/example additives | Operating principle |
|---------|---------------------------------------|---------------------|
| *Mechanical* | | |
| Liquid shear | High-pressure liquid homogenizers | Passage of cells under high pressure through a restricted orifice followed by a sudden pressure drop. Mechanisms other than shear also involved |
| Solid shear | High-pressure homogenizer | Pressure extrusion of frozen cells through a restricted orifice with ice crystals contributing to shear effect |
| Glass bead agitation | High-speed agitation of cells with glass beads | Cell disintegration achieved by abrasion |
| *Chemical* | | |
| Detergents | Tweens, sodium lauryl sulphate, sodium cholate, quaternary ammonium compounds | Modification of cell-membrane lipoproteins with release of intracellular constituents |
| Solvents | Acetone, ethyl acetate, isopropanol | Solubilization of membrane lipid material leading to cell disruption |
| Enzymatic hydrolysis | Cell's own autolytic enzymes or added lytic enzymes | Hydrolysis of cell-wall carbohydrate and protein constituents |

coat or mixed with the broth. Tangential flow (cross-flow) filtration is an effective method for separation of cells from liquid where high value products are involved. The fluid motion parallel to the membrane helps reduce the thickness of the cell layer on the filter surface (Fig. 5.7). The cell layer, rather than the membrane itself, usually exhibits the controlling resistance to the flux rate of the filtrate.

Important processes for disruption of microbial cells are summarized in Table 5.3. In most instances mechanical methods are used on a laboratory or pilot scale. Solvent and enzymatic methods are used for large-scale production of yeast extract. When recovering intracellular soluble products for further purification, cell-disruption processes should be used which minimize disintegration and solubilization of cell-wall and membrane components.

Processes for recovery of soluble products from clarified broths and cell homogenates are summarized in Table 5.4. Reverse osmosis and ultrafiltration are widely used to separate, dialyse and concentrate proteins, enzymes and hormones

**Table 5.4**  Recovery of soluble products from clarified broths and cell homogenates

| Process | *Nature of equipment/additive* | *Operating principle* |
|---|---|---|
| Size separation | Concentration by reverse osmosis | Solvent is pressure-driven through a membrane with pore size small enough to retain solutes |
| | Ultrafiltration | Use of large pore membranes such that low-molecular-weight solutes are forced through the membrane |
| | Gel filtration | Size separation using chromatographic gels with precisely controlled pore sizes. In a column the passage of smaller molecules, which penetrate the pores to a greater degree, is retarded while the larger molecules are excluded and are eluted first |
| Precipitation | Addition of chemical compounds or organic solvents which reduce the solubility of the product | Reaction of precipitant with solute to produce insoluble product, often a crystalline salt<br>'Salting out' of charged molecules<br>Enhancement of electrostatic interactions by using organic solvents to reduce medium dielectric constants<br>pH adjustments resulting in isoelectric precipitation<br>Flocculating processes |
| Adsorption | Inorganic adsorbents—charcoal, aluminium oxide, aluminium hydroxide, magnesium oxide, silica gel. Organic macroporous resins | Binding of solute to solid phase by weak van der Waals forces and possibly ionic interaction also |
| Ion-exchange absorption | Organic polymers containing reactive groups with cationic- or anionic-exchange properties | Reversible exchange of ions between liquid and solid phase, e.g. Resin–$COO^-Na^+$ + Solute $\rightleftharpoons$ Resin–$COO^-$Solute$^+$ + NaOH |
| Liquid–liquid extraction | Solvent extraction | Extraction of an aqueous medium with an immiscible organic solvent where the solute is more soluble in the organic solvent than in the aqueous phase |

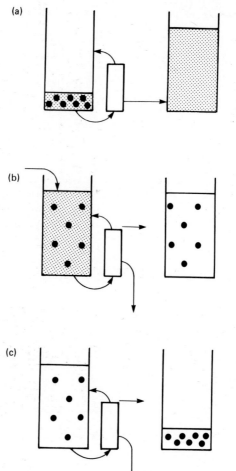

**Fig. 5.8** Separation (a), dialysis (b) and concentration (c) of proteins by ultrafiltration.

from low molecular-weight solutes (Fig. 5.8). Problems of concentration polarization, whereby an increase in solute concentration occurs adjacent to the membrane, experienced with conventional membrane systems, are reduced in cross-flow filtration systems. In cross-flow filtration the fluid motion, parallel to the filter surface, continuously removes the accumulating solute. Typical membrane systems exist as hollow fibres or flat-sheets having a plate and frame or cartridge configuration (Fig. 5.9).

Organic solvents are widely used in industry to precipitate proteins and polysaccharides and to extract antibiotics from clarified broths or cell homogenates. A variety of other separation processes exist which modify product

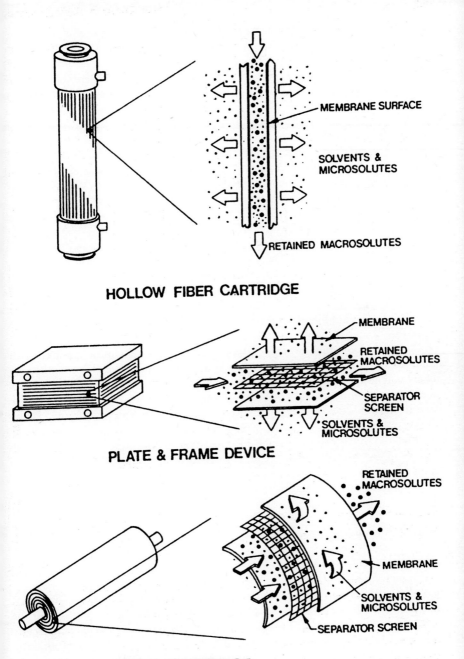

MEMBRANE SURFACE

SOLVENTS & MICROSOLUTES

RETAINED MACROSOLUTES

**HOLLOW FIBER CARTRIDGE**

MEMBRANE

RETAINED MACROSOLUTES

SEPARATOR SCREEN

SOLVENTS & MICROSOLUTES

**PLATE & FRAME DEVICE**

RETAINED MACROSOLUTES

MEMBRANE

SOLVENTS & MICROSOLUTES

SEPARATOR SCREEN

**SPIRAL CARTRIDGE**

**Fig. 5.9** Membrane configurations used in cross-flow filtration (reproduced with permission from Tutunjian, 1985).

solubility. Adsorption techniques are limited by the number of binding sites on the resin and tend to be used to bind low-volume, high-value products or to remove pigments or impurities from solute products.

More sophisticated techniques are required to isolate specific peptide or protein molecules from related peptides or proteins within the cell and sometimes also in the extracellular fluid. This is particularly the case when the application of the product, for example in clinical diagnosis or therapy, demands an extremely high level of product purity. Whereas antibiotic recovery costs account for between 40 and 60% of production costs, down-stream purification accounts for 80–90% of the costs of recombinant DNA products. Examples of recently-developed powerful preparative techniques include high-performance liquid chromatography (HPLC), electrophoresis and immunosorbent chromatography.

HPLC rapidly separates products to a high degree of purity with little or no loss of yield and was initially developed as an analytical technique. The technique involves use of strong rigid, usually polar porous particles as stationary phase and relatively non-polar solvents as mobile phase. Separations exploit small differences in polarity of molecules. Other HPLC packings have ion-exchange or molecular-sieve properties and separations based on charge or molecular-size differences may be achieved. Large-scale HPLC equipment is now available commercially, capable of processing kilograms of crude preparations in a few minutes.

The use of immunosorbent chromatography, whereby proteins and other molecules may be specifically bound to immobilized monoclonal antibodies, also offers potential for the isolation of high-value products in high yield and purified form. Indeed, monoclonal antibody packings may be used in HPLC systems to permit high-speed separations based on principles of biological affinity. Purification of monoclonal antibodies produced in either ascites or cell suspension can be hindered by the presence of non-specific globulins, introduced into the product by the mouse host or in the serum culture-medium supplement. The required monoclonal antibody may be efficiently purified using antigen-specific affinity chromatography where the antigen is immobilized to a suitable stationary phase.

## Example recovery processes

### NON-VOLATILE METABOLITES

Recovery process outlines for a number of non-volatile metabolites are illustrated in Fig. 5.10. All of these processes involve initial solid–liquid separation of the whole fermentation broth. However, riboflavin, itaconic acid and calcium lactate are first solubilized by heat treatment or pH adjustment. With the exception of gluconic acid and lactic acid which are marketed as liquid concentrates, the products illustrated are recovered in solid powder forms having crystallization and drying operations as final process stages, sometimes directly preceded by evaporation. Intermediate purification steps vary with product properties. Penicillin and steroid recoveries involve extraction operations. Advantage is

## Down-stream processing

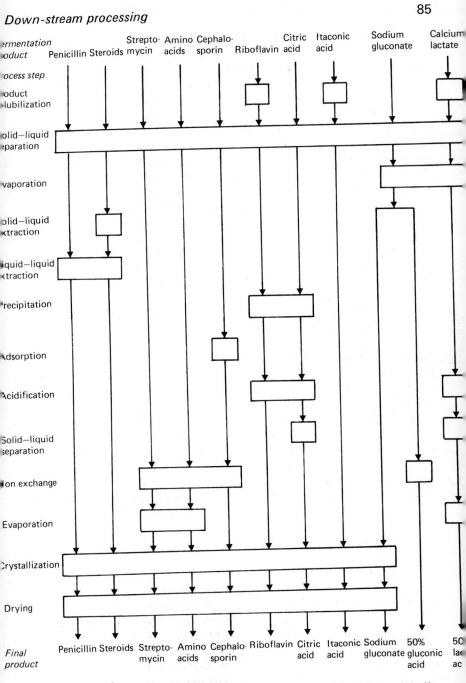

**Fig. 5.10** Process outlines for a number of non-volatile microbial metabolites.

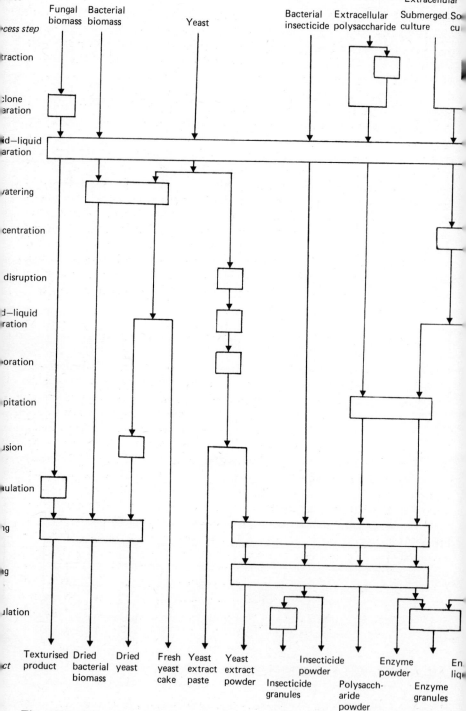

**Fig. 5.11** Recovery process outlines for some microbial biomass, extracellular polysaccharide and extracellular enzyme products.

taken of the high solubility of acid forms of Penicillin G and V in organic solvents such as amyl- and butyl-acetates in the liquid–liquid extraction step. Because of their low aqueous solubility, steroids are predominantly associated with the cell solids' fraction after broth solid–liquid separation. Recovery involves a solid–liquid extraction using an organic solvent such as acetone. This extract is then concentrated and extracted with a second solvent to effect steroid crystalliz-ation. Ion exchange is the main purification step involved in the recovery of streptomycin, amino acids and cephalosporin, preceded in the latter case by an active carbon adsorption–elution step. Ion exchange is also used to produce gluconic acid from sodium gluconate. A precipitation step is used to recover riboflavin and citrate from clarified culture broths, using a reducing agent and lime, respectively. Advantage is taken of the insoluble nature of reduced riboflavin and calcium citrate. Riboflavin is re-oxidized and dissolved using 10% hydro-chloric acid. Calcium salts of citrate and also of lactate are acidified with sulphuric acid, yielding the aqueous acids and a precipitated by-product $CaSO_4$ (gypsum) which is separated out. The solubility of itaconic acid in water, about $7 \text{ kg m}^{-3}$ at $20°C$ and $60 \text{ kg m}^{-3}$ at $80°C$, facilitates its direct crystallization from clarified culture broths. Additional process refinements including recrystallization steps, are incorporated into some of these process trains to increase product purity.

## BIOMASS, EXTRACELLULAR POLYSACCHARIDES AND ENZYMES

Example recovery processes for some microbial biomass, extracellular polysac-charide and extracellular enzyme products are illustrated in Fig. 5.11. Prior to the initial solid–liquid separation step, extracellular enzymes from surface culture are water-extracted (viscous polysaccharide broths sometimes require dilution) and spent gas is removed from fungal biomass by cyclone separation. Fungal biomass is recovered by vacuum filtration after which the mycelium is formulated with other ingredients to give a variety of final texturized products. Bacterial and yeast cells are recovered by flocculation or centrifugation, with subsequent dewatering using decanter centrifuges or other equipment. The extrusion and drying process for yeast is designed to retain optimum yeast viability. Yeast extract is produced from yeast-cell slurry usually by autolysis procedures. After removal of cell solids, the extract is evaporated or dried. Bacterial insoluble protein crystal insecticides are prepared by drying the fermenter solids' material to produce powder, granules or formulated preparations. Extracellular enzymes are recovered as cell-free con-centrates to be used as final products or precipitated using organic solvents to form enzyme powders. Dustless solid enzyme preparations are prepared by mixing enzyme powders or liquids with binding, granulating or encapsulating agents. Polysaccharides are also recovered by organic solvent precipitation.

## OTHER PRODUCTS

The processes for purification of intracellular proteins and peptides are very much

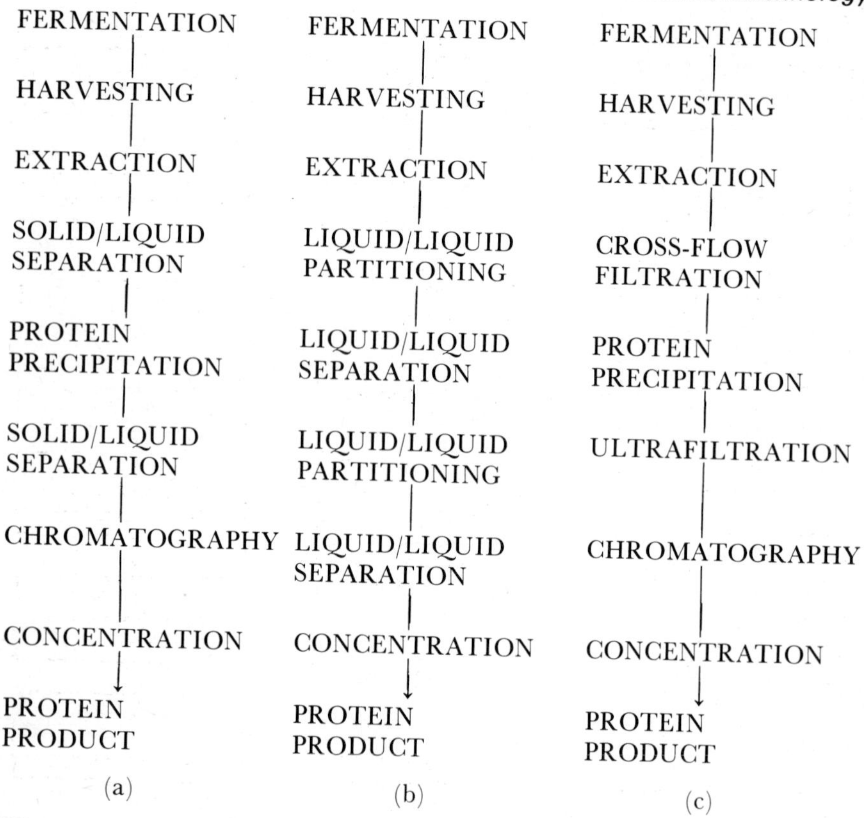

| FERMENTATION | FERMENTATION | FERMENTATION |
|:---:|:---:|:---:|
| HARVESTING | HARVESTING | HARVESTING |
| EXTRACTION | EXTRACTION | EXTRACTION |
| SOLID/LIQUID SEPARATION | LIQUID/LIQUID PARTITIONING | CROSS-FLOW FILTRATION |
| PROTEIN PRECIPITATION | LIQUID/LIQUID SEPARATION | PROTEIN PRECIPITATION |
| SOLID/LIQUID SEPARATION | LIQUID/LIQUID PARTITIONING | ULTRAFILTRATION |
| CHROMATOGRAPHY | LIQUID/LIQUID SEPARATION | CHROMATOGRAPHY |
| CONCENTRATION | CONCENTRATION | CONCENTRATION |
| PROTEIN PRODUCT | PROTEIN PRODUCT | PROTEIN PRODUCT |
| (a) | (b) | (c) |

**Fig. 5.12** Alternative purification sequences for recovery of intracellular proteins and peptides (reproduced with permission from Fish and Lilly, 1984).

dependent on the properties of the molecule and its concentration within the cell. With intracellular enzymes there is often a common sequence of purification steps consisting of extraction, nucleic acid removal, precipitation followed by one or more chromatographic steps. The more conventional purification processes tend to incorporate solid–liquid separation steps for cell harvesting, removal of cell debris and separation of precipitate (Fig. 5.12). An alternative route includes liquid–liquid extraction steps in place of solid–liquid separations and a third system uses membrane techniques for solid–liquid separation and protein purification. Three recovery processes dealing with isolation of human proteins from recombinant *E. coli* are illustrated in Fig. 5.13. The processes for recovery of largomycin F-II, a chromoprotein antitumour antibiotic produced by *Streptomyces pluricolorescens* MCRL-0367 from filtered fermentation broth and from mycelium, are illustrated in Fig. 5.14.

Other examples of recovery processes are provided in Chapters 6 to 11.

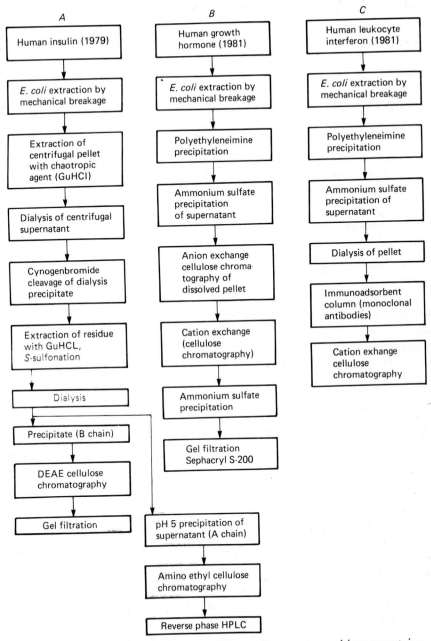

**Fig. 5.13** Recovery processes for isolation of human recombinant proteins from *Escherichia coli* (reproduced with permission from McGregor, 1983).

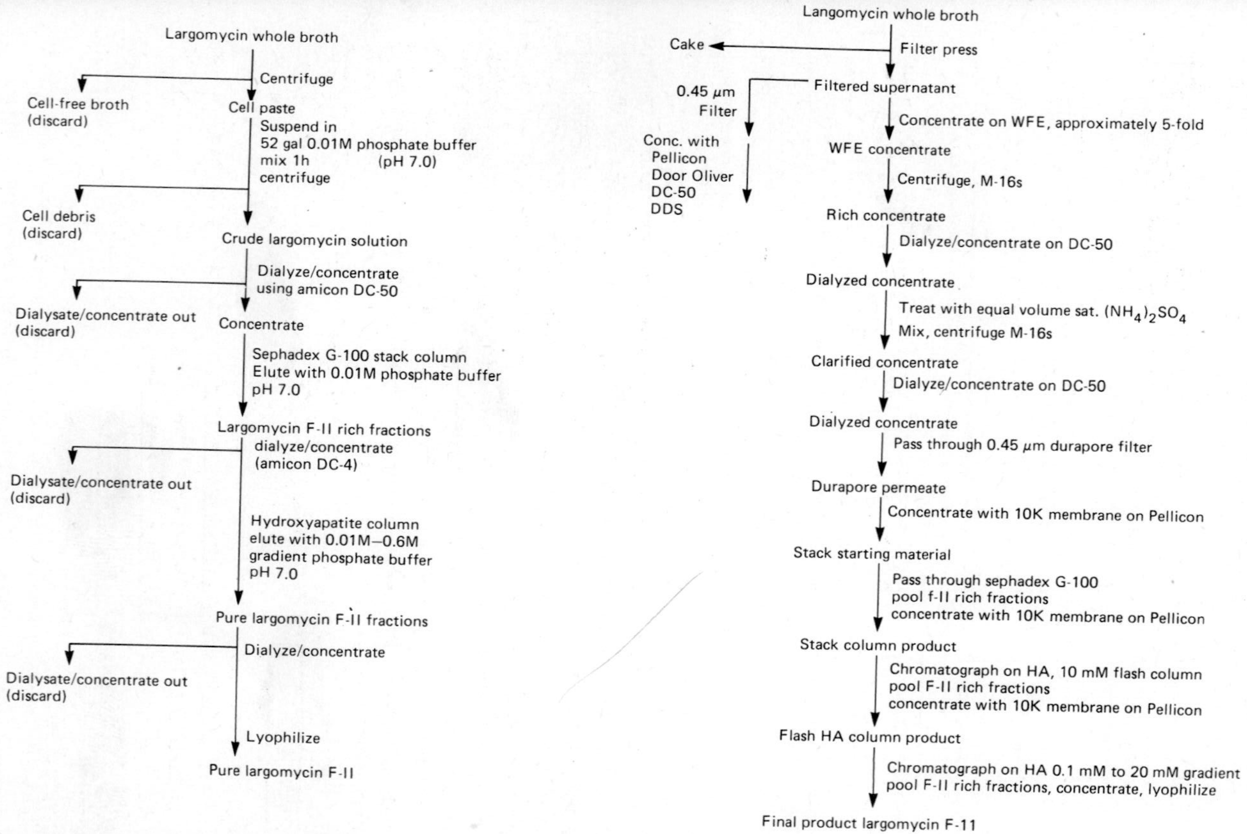

**Fig. 5.14** Processes for recovery of the antitumour antibiotic largomycin F-II from *Streptomyces pluricolorescens* (reproduced with permission from *Purification of Fermentation Products*, ACS Symposium Series 271, 1985, American Chemical Society)

# Chapter 6

# *Biomass production*

## Introduction

This chapter deals with fermentation processes involved in the production of microbial cells as major fermentation end-products. Microbial cells are produced for two main applications—as a source of protein for animal or human food (single-cell protein, SCP) or for use as a commercial inoculum in food fermentations and for agriculture, waste treatment and other applications. As a commodity, single-cell protein must be competitive with commercial animal and plant proteins in terms of price and nutritional value and must conform with human or animal food safety requirements. Productivity, yield and selling price are the major factors affecting the economics of SCP production. Microbial inoculants, which are used as a process aid, generally have a higher value. In this case the objective of the production process is to optimize yield of viable cells of defined biological activity and with good shelf-life characteristics. These two major product categories are considered separately below. *Saccharomyces cerevisiae* yeast is categorized primarily as a microbial inoculant. Inactive dried brewer's or baker's yeast is also used as a dietary source of vitamins and trace minerals in specific medical conditions. Considerable amounts of yeast extract are produced from baker's yeast as a source of flavour and vitamins.

## SCP production

### SUBSTRATES

The major substrates which have been used in SCP production are alkanes, alcohols and carbohydrates.

*Candida utilis* has been produced as a protein supplement by fermentation of sulphite waste liquor in Germany during both World Wars and by growth on molasses in Jamaica at the end of World War II. Subsequently a number of companies in the USA and one in Finland produced *Candida* as a fodder yeast on sulphite waste liquor using the German open-top Waldhof fermenter. However, due to an over-abundance of plant protein material, these processes eventually became uneconomical. More recently, in 1974 a Finnish company developed a fungal SCP production process, the Pekilo process, to grow *Paecilomyces varioti* for animal feed using spent sulphite waste liquor as substrate. A second plant was completed in 1982.

Cellulose from natural sources and waste wood is an attractive starting material for SCP production because of its abundance. The association of cellulose with lignin in wood makes it somewhat intractible to microbial degradation and thermal or chemical pretreatment, used in combination with enzymatic hydrolysis, is usually required. Systems using cellulolytic organisms appear to have promise, but economic viability has yet to be achieved.

Whole milk whey or deproteinised whey is a carbohydrate source which creates disposal problems. The major problems with whey for SCP production are usually insufficient substrate, seasonal supply variations and its high water content (>90%) which makes transport prohibitively expensive. While most organisms do not utilize lactose as a carbon source, strains of the yeast *Kluyveromyces fragilis* readily grow on lactose and a number of fermentation plants have been built to produce animal or human food-grade SCP using this organism. Some of these plants are designed to switch from SCP production to ethanol production depending on market forces.

The Symba process was developed in Sweden to produce SCP from potato starch using two yeast strains. *Saccharomycopsis fibuligera* produces the enzymes necessary for starch degradation enabling co-growth of *Candida utilis*. The production process, intended for animal feed production was designed to handle potato-processing waste but suffered from problems of lack of substrate supply continuity. Food-grade glucose was the substrate chosen by Rank Hovis McDougall for production of fungal SCP using *Fusarium graminearum*. The strategy adopted was to additionally take advantage of mycelial fibre content to produce a range of high added value products including meat analogues for human consumption.

The original alkane SCP fermentation process, developed by BP at the Laverna Refinery in France, used 10–20% wax contained in gas-oil. Substrate costs were very low but, because of its crude nature, exhaustive extractive processing was required to recover the yeast free of a gas-oil flavour taint and possible gas-oil carcinogens. There was also a microbial contamination tendency due to the non-aseptic nature of the process. These drawbacks led to the closure of the plant in 1975. The disadvantages associated with use of crude gas-oil and non-asepsis were taken into account by Italproteine, a joint venture between BP and ANIC. The process used purified *n*-paraffins which were fully utilized by the cells, simplifying SCP recovery. However, the 100 000 t/year capacity SCP plant, which was constructed for the process in Sardinia in 1976, was never allowed to operate commercially. Japanese companies, Dainippon and Kanegufuchi, working along similar lines to Italproteine, gained clearance in Japan to have their products accepted for animal feed but the decision was reversed by a strong consumer campaign based on fears of the presence of carcinogenic residues. The Kanegufuchi process was licensed to Liquichimica in Italy, who built a 100 000 t/year capacity plant which was then subjected to an embargo similar to the Sardinia plant. A plant, designed by Dainippon, in Romania using *Candida pichia* was reported to have been commissioned.

Methane was initially considered as an SCP raw material because, as a gas product, purification problems after fermentation would be minimal. Disadvantages associated with methane-based processes are related to the greater oxygen requirement necessary to fully oxidize methane compared to paraffins, the low solubility of methane in water and the requirement that the fermentation plant be flame-proof as methane–oxygen mixtures are highly explosive. Methane is, however, easily converted to methanol which requires less oxygen, less fermenter cooling, is highly water-soluble and has minimal explosion risks. ICI, which manufactures bulk methanol, chose this substrate for bacterial SCP production for animal feed. The company designed a non-mechanical 'pressure cycle fermenter', which uses air for both agitation and aeration, in the world's largest single aerobic aseptic fermenter of 3000 m$^3$ capacity. The process, which produces 50–60 000 t/year SCP, using the organism *Methylophilus methylotrophus*, was commissioned in 1979–80 but has suffered from dramatic increases in methanol prices. The economic difficulties encountered by ICI with the animal-feed process and the greater commercial promise of the RHM *Fusarium* process led to a joint venture between these companies in 1983 with the objective of producing *Fusarium* SCP in the large ICI fermentation plant. Ethanol, which has similar advantages as a substrate to methanol, was used as substrate by Pure Culture Products for production of food-grade protein from food-grade ethanol using *Candida utilis* in 1975. Process economics likewise suffered as a result of increases in ethanol costs.

## ECONOMICS OF SCP PRODUCTION

The initial reasoning by companies such as BP and ICI to enter SCP production was to produce, at low cost, high-value SCP from petroleum, for addition to animal feed, thereby replacing imported protein additives such as soyabean meal. Factors which contributed to the failure of hydrocarbon SCP to make a major commercial impact include the 1973 dramatic oil price increases which raised feedstock and energy costs, parallel increases in plant construction costs and the lower price increases achieved by agricultural products including soyabean relative to industrial products. When one considers that crude oil prices increased by a factor of six in 1973 and that the cost of substrate for SCP processes represents 40–60% of total manufacturing costs, the negative impact on hydrocarbon-based SCP processes may be understood. Agricultural crops, the major competitor to SCP for animal feed, manifest a remarkable ability to respond to market forces and maintain price stability. In addition to the more conventional animal-feed crops such as soya, high-protein crops such as ground nut, rape seed, cotton seed and winged bean are now gaining a market share in this business and expansion of maize gasohol production has provided new sources of animal-feed by-products. Consequently, the economics of animal-feed SCP have so far appeared unattractive and higher value products are being sought. Rank Hovis McDougall and Pure Culture Products have adopted a different strategy and aimed their products at the human food market. RHM in particular have taken advantage of the fibre component of the fungus to produce high added value meat analogues containing high dietary fibre, 50% protein and having other desirable advantages such as low sodium and low fat.

The main economic factors in SCP production are productivity, yield and selling price. Table 6.1 summarizes productivity and yield factors for SCP production using various organism/substrate combinations. Cell dry weight and dilution factors are usually $15-30 \, \text{kg m}^{-3}$ and $0.1-0.4 \, \text{h}^{-1}$ respectively giving productivity (weight of cells produced per unit volume per unit time) of $1.5-12 \, \text{kg m}^{-3} \, \text{h}^{-1}$. Since productivity is often limited by oxygen-transfer rates and effectiveness of fermenter cooling due to exothermic metabolism, values in the range $3-5 \, \text{kg m}^{-3} \, \text{h}^{-1}$ are more normal. Because substrate cost is such a large proportion of manufacturing cost of most SCP products, high cell yield (weight of cells produced per unit weight of substrate utilized) and minimal formation of by-products is essential.

## CHOICE OF MICRO-ORGANISM

Key criteria used in selecting suitable strains for SCP production should take into account the following:

(1) The substrates to be used as carbon, energy and nitrogen source and the need for nutrient supplement.
(2) High specific growth rates, productivity and yields on a given substrate.
(3) pH and temperature tolerance.
(4) Aeration requirements and foaming characteristics.
(5) Growth morphology in the fermenter.
(6) Safety and acceptability—non-pathogenic, absence of toxin products.
(7) Ease of SCP recovery.
(8) Protein, RNA and nutritional composition of product.
(9) Structural properties of the final product.

In general, fungi have the capacity to degrade a wider range of complex plant materials, particularly plant polysaccharides. They can tolerate low pH which contributes to reducing fermenter infections. Growth of fungi as short highly-branched filaments rather than in pelleted form is essential in order to optimize growth rate. However, this filamentous morphology produces rheologically more complex fermentation broths which are difficult to aerate.

Bacteria in general have faster growth rates than fungi and grow at higher

**Table 6.1** Productivity and yield factors of SCP biomass, exfermenter, achieved in pilot or production scale plants (reproduced with permission from Solomons, 1983)

| Substrate | Organism | Yield ($Y$) (kg cell wt/kg substrate utilized) | Cell dry wt ($x$) (kg m$^{-3}$) | Dilution rate (h$^{-1}$) | Productivity (kg m$^{-3}$ h$^{-1}$) |
|---|---|---|---|---|---|
| n-Paraffin | Yeast | 0.95 | 15–20 | 0.11 | ca. 2 |
| n-Paraffin | Yeast | 1.2 | | | 3 |
| Methanol | Bacteria | 0.4 | | | 2 |
| Ethanol | Yeast | 0.8 | | | 4.5 |
| Methanol | Yeast | 0.85 | | | 5.2 |
| Molasses | Yeast | 0.5 | 20–25 | | 8–10 |
| Methanol | Bacteria | ca. 0.5 | | 0.4 | 2.8–3.4 |
| Sulfite waste liquor | Fungus | | 17 | 0.2 | |
| Methanol | Bacteria | ca. 0.5 | 30 | 0.16–0.19 | 4.8–5.7 |

**Table 6.2** Composition (%) of single-cell protein compared with soya meal and milk powder

| Component | Alkane yeast | Methanol bacterium | Fusarium graminearum* | Alga | Soya meal | Milk powder |
|---|---|---|---|---|---|---|
| Raw protein | 60.0 | 80.0 | 44.3 | 72.6 | 42.0 | 34.0 |
| Fat | 9.0 | 9.5 | 13.8 | 7.3 | 4.0 | 1.0 |
| Nucleic acid | 5.0 | 15.0 | 3.1 | 4.7 | 6.5 | 8.0 |
| Mineral salts | 6.0 | 9.5 | | | 40.0 | |
| Amino acids | 54.0 | 65.0 | | | | |
| Moisture | 4.5 | 2.8 | 0 | 3.6 | 10.0 | 5.0 |

*after RNA reduction

temperatures, thereby reducing fermenter cooling requirements. Bacterial and yeast fermentations are easier to aerate. In contrast to fungi, which are easily recovered by filtration, bacteria and yeasts require use of sedimentation techniques including centrifugation.

Bacteria in general produce a more favourable protein composition than yeasts or fungi. Protein content in bacteria can range from 60 to 65% whereas fungi selected for biomass production and yeasts have protein contents in the range 33 to 45%. However, associated with the higher bacterial protein levels is a much higher level of nutritionally undesirable RNA of 15 to 25%.

Micro-organisms involved in SCP production must be safe and acceptable for use in food or feed. They should be non-pathogenic and non-toxin forming. Organisms should be stable genetically so that the strain with optimal biochemical and physiological characteristics may be maintained in the process through many hundreds of generations. In addition, regulatory bodies are concerned that strain degeneration could result in production of a strain with undesirable nutritional characteristics.

## FERMENTER DESIGN

Economics dictates that production should be carried out in the minimum number of large-scale fermenters. Key parameters in the achievement of high biomass productivity on a large scale are high oxygen-transfer rates, promoting high respiration rates which in turn increase metabolic heat production and the need for an efficient cooling system. Assuming a mass balance of biomass-producing organisms of

$$C_6H_{12}O_6 + O_2 + NPKMgS \rightarrow Biomass + CO_2 + H_2O$$
$$(2.0) \quad (0.7) \quad (0.1) \quad\quad (1.0) \quad (1.1) \quad (0.7)$$

and a heat evolution of 3–4 kcal g$^{-1}$ cell mass, a cell-mass productivity of 4 kg m$^{-3}$ h$^{-1}$ requires an OTR of 2.8 kg m$^{-3}$ h$^{-1}$ and produces 14 000 kcal m$^3$ h$^{-1}$ heat. In order to maximize fermentation productivity it is essential to operate continuous fermentation processes, thereby maintaining high microbial growth rates and minimizing fermenter down-time.

Some of the fermenter designs used in SCP production are illustrated in Fig. 6.1. Mechanically-agitated fully-baffled bioreactors (Fig. 6.1a) with turbine mixers and with air introduction through a sparger have been used by BP for the *n*-alkane pilot-scale process developed in Scotland and for the three 1800 m$^3$ fermenters in Sardinia. Conventional turbine systems are not considered to be very satisfactory for very large fermenters and a number of large scale SCP processes have used air-agitated vessels. For gas-oil substances, BP used a draft tube air lift design (Fig. 6.1b). A correlation existed between the rising speed of air bubbles in the draft tube and fermenter productivity and the draft tube was optimized relative to biomass while minimizing energy requirements. A Kanegufuchi-designed, modified air-lift fermenter, in which the fermentation medium is driven by the force of inflowing air from a large vessel through an external circulatory loop, was used for the Liquichimica alkane plant (Fig. 6.1c).

*Biomass production*

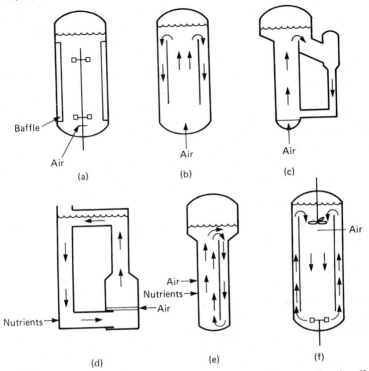

Baffle

Air

(a)

Air

(b)

Air

(c)

Nutrients

(d)

Air
Nutrients
Air

(e)

Air

(f)

**Fig. 6.1** Fermenter designs used in SCP production. See text for details.

The ICI pressure cycle pilot fermenter, used in SCP production from methanol, is likewise a combination of an air-lift and loop reactor consisting of an air-lift column, a down-flow tube with heat removal and a gas-release space (Fig. 6.1d). The production fermenter no longer contained the external down-flow tube (Fig. 6.1e). This fermenter is equipped with a complex air sparger containing 3000 outlets, which facilitates aeration, agitation and effective distribution of methanol substrate, which is toxic in high concentrations.

While air-lift fermenter designs may be used for fungal pelleted growth, only filamentous growth is appropriate for fungal SCP production. Long and/or highly branched filamentous mycelia, even at low cell densities ($< 10 \, kg \, m^{-3}$) tend to develop high pseudoplastic viscosities resulting in inefficient oxygen transfer. Pellets also have 'inefficient' internal oxygen transfer. In addition there is a complex relationship between impeller shear and cell morphology. Low shear rates induce long unbranched mycelia, having few growth tips and low growth rates. In order to optimize mass transfer in fungal biomass fermentations involving *Fusarium graminearum*, a fermenter was designed, which separates these functions, with two impellers each driven by a separate shaft and run at optimum speeds (Fig. 6.1f). Following the development of a joint venture between RHM and ICI, the ICI pilot SCP fermenter was modified to produce the RHM 'Mycoprotein' but difficulties were encountered in scaling up the *Fusarium* process in the ICI production fermenter. Nevertheless, the outlook for the process is optimistic.

## PRODUCT QUALITY AND SAFETY

SCP has potential applications in animal feed, human food and as functional protein concentrates. Nutritional characteristics are important in the case of feed and food and safety considerations are relevant to all these applications. The overall composition of selected single-cell protein preparations is illustrated in Table 6.2.

Amino acid values for SCP may be compared with FAO reference protein. In this comparison some bacterial preparations have amino acid profiles, including methionine content, that compare favourably with FAO values. Yeast, fungal and soy bean proteins tend to be deficient in methionine. Nutritional value of SCP may be determined by performance in animal control feeding tests. Evaluation methods are based on determination of the coefficient of digestibility, net protein utilization (NPU), nitrogen balance studies and protein efficiency ratio (PER).

Functional quality may relate to properties such as water- and fat-binding effectiveness, emulsion stability, dispersability, gel formation, whipability and thickening or to textural properties of whole cells or mycelia.

Ingestion of RNA from non-conventional human foods should be limited to 2 oz per day. Ingestion of purine compounds, arising from RNA breakdown, leads to increased plasma levels of uric acid which may cause metabolic disturbances in man and some primates, such as gout and kidney stone formation. The high content of nucleic acids causes no problems to animals, since uric acid is converted to allantoin which is readily excreted in urine. Consequently, there is no need for nucleic acid removal from biomass for animal feeds but in the case of human foods, nucleic acid removal is essential. Alkali treatments have been recommended in the past but can result in the production of lysinoalanine, a nephrotoxic factor. More recent methods rely on temperature holds around 64°C, which inactivates fungal proteases and allows endogenous RNA-ases to hydrolyse RNA with release of nucleotides from cell to culture broth. A 30-min stand in a continuous stirred tank reactor at 64°C reduces RNA levels in *F. graminearum* cells from about 80 mg g$^{-1}$ to 2 mg g$^{-1}$.

## SCP FERMENTATION PROCESSES

In this section fermentation processes for production of SCP by *Candida* species grown on alkanes, by *Methylophilus methylotrophus* grown on methanol, by *Kluyveromyces fragilis* grown on whey and by *Fusarium graminearum* grown on glucose, will be discussed.

The *Candida* BP *n*-paraffin process flow sheet is illustrated in Fig. 6.2. The fermentation is operated under sterile conditions. Approximate nutrient requirements to produce 1 kg of SCP included 1–1.2 kg paraffin, 0.14 kg gaseous $NH_3$, 0.05 kg $PO_4^{-3}$ plus other salts. Gaseous ammonia was fed, with air, both as nitrogen source and to control pH. Oxygen requirement per unit biomass

*Biomass production*

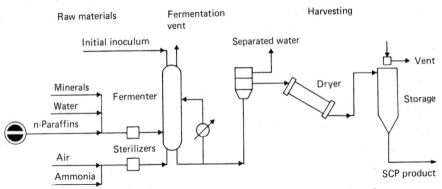

**Fig. 6.2** BP *n*-paraffin SCP process (reproduced with permission from *Hydrocarbon Processing*, November, 1974).

produced by aerobic micro-organisms grown on *n*-hexadecane is 2.5 times higher than that required for growth on glucose and amounts to 2.2 kg $O_2$ per kg biomass. Heat produced was 6600 kcal/kg biomass. Consequently fermenters require substantial agitation. Because of the insoluble nature of alkanes, they exist in agitated fermenter broths as suspensions of alkane drops 1–100 $\mu$m in size. Hydrocarbon assimilation into cells appears to involve cell contact with tiny hydrocarbon droplets, 0.01–0.5 $\mu$m diameter. Creation of a micro-emulsion at the interface of the tiny droplets by a surface-active agent, which may be produced by the cells, seems to facilitate droplet adherence to the cell. The chief mechanism of alkane degradation in *Candida* is terminal oxidation, with sequential production of primary alcohol, aldehyde and acid, followed by $\beta$-oxidation of the fatty acid to acetate. Cell recovery is by centrifugation, producing 15% dry solids, evaporation to 25% dry solids and spray-drying.

The process outline for production of SCP from methanol by *Methylophilus methylotrophus* is illustrated in Fig. 6.3. The fermentation is run aseptically in the

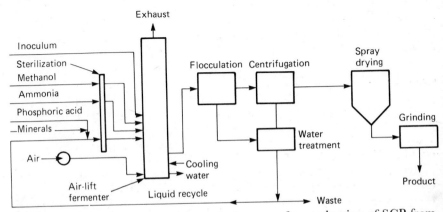

**Fig. 6.3** Schematic diagram of a typical process for production of SCP from methanol (reproduced with permission from Litchfield, 1983).

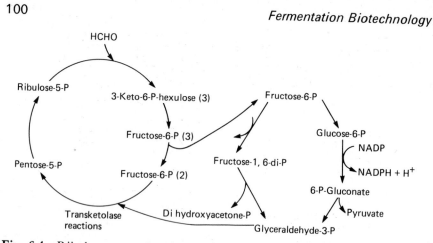

**Fig. 6.4**  Ribulose monophosphate cycle.

ICI pressure cycle fermenter. The nitrogen source is ammonia gas and pH is controlled between 6.0 and 7.0. Cell-specific growth rate is approximately $0.5\,h^{-1}$ and cell yield 0.5. Methanol is oxidized via dehydrogenation to formaldehyde which can either be assimilated for conversion to cell mass or further oxidized to $CO_2$ with concomitant energy production. In *Methylophilus*, formaldehyde is assimilated by the ribulose monophosphate pathway with ultimate conversion to fructose-6-phosphate (Fig. 6.4). Cells are recovered by agglomeration followed by

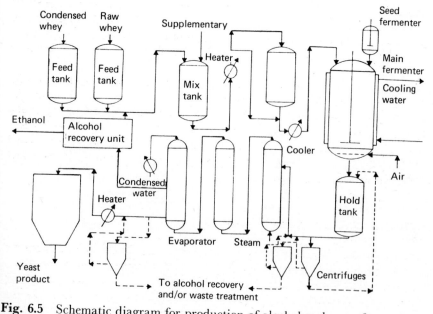

**Fig. 6.5**  Schematic diagram for production of alcohol and yeast from whey (reproduced with permission from Bernstein *et al.*, 1977).

# Biomass production

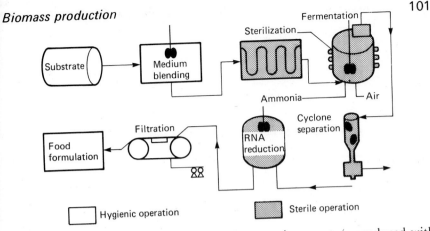

**Fig. 6.6** The Rank Hovis McDougall Mycoprotein process (reproduced with permission from Anderson and Solomons, 1984).

centrifugation, flash-dried and ground. A feature of the process is the recycle of fermenter water.

The process outline for production of biomass or alcohol by *K. fragilis* grown on whey is illustrated in Fig. 6.5. Whey, which contains about 5% lactose, 0.8% protein, 0.7% minerals and 0.2–0.6% lactic acid, may require supplementation with biotin. Biomass production requires an aerobic fermentation whereas aeration is minimal for ethanol production. For feed-grade biomass, the entire fermentation contents, containing yeast, residual whey proteins, minerals and lactic acid may be recovered. For preparation of food-grade material, cells are harvested by centrifugation, washed and dried. Cell yield is 0.45–0.55 based on lactose consumed.

The RHM Mycoprotein fermentation process flow diagram is illustrated in Fig. 6.6. Medium constituents include food-grade glucose syrup, gaseous ammonia, salts and biotin. Fermentation pH is controlled at 6.0 by gaseous ammonia addition, fed into the air inlet stream. Cell concentrations are 15–20 kg m$^{-3}$ and a specific growth rate of up to 0.2 h$^{-1}$ is achieved. Following cyclone separation and an RNA reduction step, cells are recovered by rotary vacuum filtration and formulated into a range of products.

## PHOTOSYNTHETIC SCP PRODUCTION

All SCP processes discussed so far are based on recycling of reduced organic matter and this appears to limit processes to relatively high-cost, low-volume products. Conventional agriculture has a low photosynthetic efficiency, which only stores about 1% of available solar energy. In processes which are at the interface between traditional agriculture and modern biomass production, phototrophic organisms have been cultivated in large lagoons and algae are used as part of the

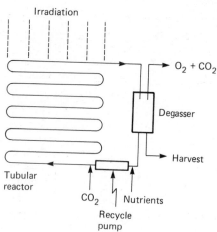

**Fig. 6.7** Diagram of tubular-loop reactor for single-cell photosynthesis (reproduced with permission from Pirt, 1982).

human diet by the Aztecs of Central America and by the natives of the Lake Chad area of Africa. Large-scale tubular loop bioreactors with fully-controlled continuous culture systems for photosynthetic cell cultivation could store up to 18% of solar energy and produce cell densities in excess of $20 \, \text{g} \, \text{l}^{-1}$ dry weight (Fig. 6.7). It has been argued that suitable photosynthetic bioreactors would permit a solution to the global problems of energy, food and chemical feedstock supplies.

## Microbial inoculants

### FOOD STARTER CULTURES

Starter cultures are used in modern food processing to accomplish various food property changes such as texture modification, preservation, flavour development or nutritional improvement. Food starter culture applications are designed to achieve these effects in a manner which complies with the requirements of process automation, product quality and reproducibility. This involves production of cultures of defined activity in terms of viability, effectiveness and shelf life. Major applications of food starter cultures exist in the baking and dairy industries. Commercial yeast starter cultures are also available for use in alcoholic beverage fermentations and industrial alcohol production. Processes for production of the latter starters are similar to those used for manufacture of baker's yeast and indeed baker's yeast is often used in alcoholic fermentations. This section is confined to discussions on baker's yeast, dairy and meat starter cultures.

### Baker's yeast

Baker's yeast, *Saccharomyces cerevisiae*, used in bread-making, degrades dough sugars into a mixture of alcohol and carbon dioxide gas bubbles which become fixed in the dough. By excretion of compounds such as cysteine and glutathione, which breaks intramolecular disulphide bonds and by gas evolution, yeast acts to chemically modify and mechanically stretch gluten, the main protein of wheat.

Depending on environmental conditions, baker's yeast fermentations can be manipulated to favour alcohol production or biomass production. Under anaerobic conditions alcohol production is optimized. Maximum theoretical biomass yield coefficient under these conditions is $Y_s = 0.075$. Under aerobic conditions in the presence of high sugar concentrations, a substantial level of alcohol is also formed due to the Crabtree effect. Maintenance of low sugar concentrations in aerated yeast fermentations by continuous sugar feeding favours biomass production. Maximum theoretical biomass yield coefficient $Y_s = 0.54$. When excess oxygen is present, a yeast culture growing in 1.1 mM glucose $(0.2\,g\,l^{-1})$ exhibits a maximum respiration rate $(Q_{O_2})$, a specific growth rate of $\mu = 0.2$–$0.25$ and negligible ethanol production. Under these conditions the respiratory quotient $(RQ = Q_{CO_2}/Q_{O_2})$ value is about 1.0. At increased glucose levels, respiration or oxygen transfer diminishes, ethanol is produced and the RQ value increases.

Molasses is the most widely used substrate for baker's yeast production. Molasses is generally deficient in nitrogen and phosphorus with respect to yeast nutritional requirements and needs to be supplemented with ammonium salts, orthophosphoric acid or other suitable phosphate forms. Biotin, which is required for growth of baker's yeast, is present in sufficient amounts in cane molasses but has to be added to beet molasses. Alternatively, a blend of beet with at least 20% cane molasses may be used. Urea may replace ammonium salts as nitrogen source provided sufficient biotin, which is also required for urea hydrolysis, is incorporated in the medium.

Yeast production is a batch process involving up to eight scale-up stages. Typical production fermenter volumes are 50–350 m$^3$ or more. The first two inoculum stages usually involve aseptic fermentations. Later inoculum and production stages are usually not aseptic and do not involve pressure vessels. The final inoculum development stage and the main biomass production stage involve incremental molasses feeding. The principal objective of the main production stage is to achieve a high viable yield of yeast with an optimal balance of properties, high fermenting activity and good storage quality. Storage quality is improved by allowing some aeration to continue after molasses feeding is stopped which improves yeast uniformity and reduces RNA and protein synthesis.

In the production fermenter the aerobic fermentation is carried out at 28–30°C and requires cooling since 3.5 kcal are generated per gram of yeast solids produced. The pH values range from 4.1 to 5.0. The pH values in the lower end of this range reduce bacterial contamination but tend to increase adsorption of molasses-colouring material. Most commercial fermentation processes start at lower pH values (4.0–4.4) and end at a higher pH range (4.8–5.0). Yeast concentrations of 40–60 g l$^{-1}$ (dry) are obtained.

The amount of oxygen required for yeast growth is approximately 1 g or 31 mM per gram of yeast solids. At the later stages of a yeast batch fermentation, yeast may be produced at a rate of up to 5 g yeast solids per litre per hour, so that the fermenter aeration system must have an oxygen-transfer capability of approximately 150 mM per litre per hour. In general, efficient gas-sparged systems are used without impellers.

Solid–liquid separation is by continuous centrifugation (nozzle-type) to produce a yeast cream of 18–20%, which is further concentrated to 27–28% solids by rotary vacuum filtration during which starch may be used as a filter aid. Higher solids levels may be achieved by salting the cream prior to filtration, which reduces cellular moisture content osmotically. The crumbly cake produced by filtration is a widely marketed final product. Emulsifiers may be blended with the yeast cake to facilitate extrusion. Active dried yeast is usually produced using fluidized bed driers, whereby heated air is blown upwards through yeast particles at velocities capable of suspending the yeast in a fluid bed. For rapid drying cycles of 10 and 30 min, air temperatures of 100–150°C are used at the start of the cycle while keeping the yeast temperature at 24–40°C. The final moisture content is 7%.

*Starter cultures for the dairy industry*

Micro-organisms are used in the manufacture of fermented dairy products to produce lactic acid, to secrete metabolites of characteristic flavour and to achieve other desirable chemical changes. The most important bacterial species involved as cheese-starter cultures are the mesophilic species *Streptococcus cremoris* and *Streptococcus lactis*, which are homofermentative, producing only lactic acid. For yoghurt manufacture the inoculum consists mainly of two thermophilic homofermentative starters, *Streptococcus thermophilus* and *Lactobacillus bulgaricus*, sometimes with *Lactobacillus acidophilus* incorporated at low levels. White moulds, *Penicillium camemberti* and its biotypes, *Penicillium candidum* and *Penicillium caseioculum*, are used in manufacture of surface mould-ripened cheese like Brie and Camembert and blue moulds such as *Penicillium roqueforti* are used in internal mould ripened cheeses such as Gorgonzola and Stilton.

Technology for fungal-spore production is relatively straightforward involving small-scale surface-culture methods. Spores are produced in powdered form by commercial culture firms.

Both bacterial mesophilic starter species are susceptible to invasion by bacteriophages, the major cause of starter failure in cheese-making and consequently the major thrust of starter-culture development is directed at overcoming problems related to bacteriophage infection. During the 1930s and 1940s crude mixed cheese-starter cultures were maintained by sub-culturing for years. Intensification of cheese-making through accelerated demand increased susceptibility of these crude cultivation practices to failure caused by bacteriophage. Single starter strains were introduced into commercial cheese-making and variability in starter activity became a serious problem. Measures taken to control bacteriophage contamination included aseptic starter propagation, sanitation of cheese-making equipment, use of enclosed cheese-making vats and shortening of ripening times. These practices were all insufficient to resolve the bacteriophage problem. From the 1950s, commercial culture suppliers provided freeze-dried and later frozen defined or undefined starter culture mixtures to cheese factories for

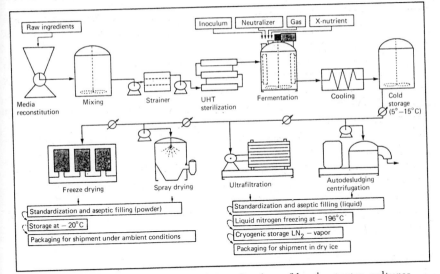

**Fig. 6.8**   Flow scheme for large-scale production of lactic starter cultures (reproduced with permission from Porubscan and Sellars, 1979).

both 'mother' or bulk starter preparation. Most of the *S. cremoris* and *S. lactis* strains were only sensitive to a specific number of phage types, and since sensitivity and resistance patterns differed among strains of an individual species, phage levels could be controlled by frequent rotation of phage-unrelated strains. Introduction of phage inhibitory media, containing phosphate salts to chelate the calcium required for bacteriophage propagation, minimized phage build-up during bulk starter preparation. Commercial concentrated cultures were introduced for direct inoculation of bulk starter tanks and later for direct cheese vat inoculation. Acceptable commercial dairy starter strains must be effective in acid and flavour production. Since the beginning of the 1980s, bacteriophage-insensitive strains have been successfully developed.

A flow diagram for large-scale production of lactic starter cultures is illustrated in Fig. 6.8. The objective of the fermentation and recovery process is to achieve a high yield of bacteriophage-free starter culture product. Fermentation media usually contain complex nutrients such as milk, whey, yeast extract, peptones, mono- or di-saccharides, vitamins, buffers, salts and phage-inhibiting constituents. Media may be sterilized by UHT processes. Optimum temperature for growth will vary depending on strain but is generally in the region 25–30°C for streptococci and 30–37°C for lactobacilli. pH control is desirable and generally in the range pH 5.4–6.3, with pH 6.0 preferred, using ammonia gas, other alkalis or internal buffering. The fermentation is usually agitated by gentle stirring. After the fermentation is complete, the broth is cooled and may be concentrated aseptically by filtration to yield a culture concentrate for direct inoculation or for storage by freezing or freeze-drying.

**Table 6.3** Other applications of microbial inoculants

| Applications | Microbial species | Function |
|---|---|---|
| Insecticides | *Bacillus thuringiensis* | Pathogenic to lepidoptera larvae |
| | *Bacillus popilliae* | Pathogenic to Japanese beetle |
| | *Bacillus alvei,* | |
| | *Bacillus circulans* and | Pathogenic to mosquitoes |
| | *Bacillus sphaericus* | |
| | *Serratia piscatorum,* | |
| | *Streptococcus faecalis* and | Kill butterfly larvae by reducing intestinal pH |
| | *Aerobacter aerogenes* | |
| | *Beauveria tenella* | Pathogenic to May bug larvae |
| | *Veticillium lecanii* | Controls aphids and white flies |
| | *Hirsutella thompsonii* | Controls mites on citrus plants |
| | *Metarrhizium* sp. | Controls lepidoptera species |
| Mineral cycling | *Thiobacillus* spp. | Sulfur oxidation |
| | *Beggiatoa* spp. | Oxidizes $H_2S$ in rice soils |
| | *Bacillus megaterium,* | |
| | *Bacillus, Pseudomonas* | Phosphate mineralization and phosphorus dissolution |
| | and *Chromobacterium* spp. | |
| Nitrogen fixation | Free-living and symbiotic nitrogen-fixing prokaryotes | Emphasis on production of legume inoculants of *Rhizobium* species to increase nitrogen fixation |
| Plant growth acceleration using mycorrhizal fungi | Endomycorrhizae of the family Endogonaceae, phylum Zygomycota (vesicular-arbuscular fungi). Ectomycorrhizae of the phylum Dikaryomycota-majority are Basidiomycetes | Increased root nutrient and water uptake, protection against disease |
| Silage-making | *Streptococcus faecium, Lactobacillus plantarum,* other acid-producing bacteria | pH reduction to accelerate ensilation |
| Probiotics | *Lactobacillus acidophilus, Steptococcus faecium,* rumen bacteria | Increase digestibility of animal feeds |
| Waste treatment | Methanogens | Anaerobic digestion |
| | *Pseudomonas* spp., *Acinetobacter* spp. and *Nocardia* spp. | Degrade alkanes and recalcitrant waste compounds |
| | Extracellular enzyme-producing bacteria | Hydrolyse waste proteins, carbohydrates and fats |

*Biomass production*

*Meat starter cultures*

The primary objective of fermenting meats is for preservation purposes. A microbial fermentation results in acid development which causes moisture removal, provides a desirable safety margin, improves stability and produces quality-related properties such as flavours and aromas which have not been duplicated solely by use of chemical additives. The majority of fermented meat products consist of dry and semi-dry sausages. Technical improvements in sausage-making were achieved by use of added sugar, to encourage rapid production of lactic acid, and the use of nitrate, which lowers the oxidation–reduction potential of the meat as nitrate is converted to nitrite. This results in stabilization of meat colour by preventing haemoglobin oxidation, provides an environment which favours micro-aerophilic lactic acid producers and suppresses development of undesirable bacteria. Desirable characteristics of meat starter cultures are (a) their ability to produce lactic acid (b) their tolerance to salts, spices, nitrate and nitrite and (c) their capacity to reduce nitrate. More recently, the industry has moved towards using nitrite rather than nitrate, thereby eliminating the requirement for nitrate-reducing starters in some meat fermentations.

Organisms which have been successfully exploited as meat starter culture inoculants are (i) *Pediococcus cerevisiae*, which has acid-producing properties but does not reduce nitrate, (ii) *Micrococcus* species which reduce nitrate and contain catalase to remove any peroxides produced during the meat fermentation and (iii) other acid-producing bacteria such as *Lactobacillus plantarum*, *Lactobacillus brevis* and *Leuconostoc mesenteroides*. Single and mixed-strain variants of these cultures are available commercially in frozen and freeze-dried form and it is predicted that the meat industry will make much greater use of microbial inoculants in the future not only in fermented meat processes but also to control growth of undesirable micro-organisms.

OTHER APPLICATIONS OF MICROBIAL INOCULANTS

Microbial inoculants are used in a variety of non-food applications. Some key existing or potential applications are listed in Table 6.3.

# Chapter 7

# *Food fermentations*

Micro-organisms have been used for centuries to modify foodstuffs and fermented foods and beverages constitute a major and extremely important sector of the food industry. This chapter will describe some of the applications of fermentation processes in production of alcoholic beverages, cheese-making, bread-making, fermented soya-based foods, meat processing and vinegar manufacture.

## Alcoholic beverages

Alcoholic beverages are produced from a range of raw materials but especially from cereals, fruits and sugar crops. They include non-distilled beverages such as beers, wines, ciders, and sake. Distilled beverages such as whisky and rum are produced from fermented cereals and molasses, respectively, while brandy is produced by distillation of wine. Other distilled beverages, such as vodka and gin, are produced from neutral spirits obtained by distillation of fermented mollases, grain, potato or whey. A variety of fortified wines are produced by addition of distilled spirit to wines to raise the alcohol content to 15–20%. Notable products include sherries, Port and Madeira wines.

   An important common feature in the production of all these alcoholic beverages is the use of yeast to convert sugars to ethanol. Consequently, the first part of this section will concentrate on the biology of yeast fermentations. Processes for production of beer, whisky and wine will then be discussed.

### BIOLOGY OF YEAST FERMENTATIONS

Over 96% of fermentation ethanol is produced using strains of *Saccharomyces cerevisiae* or species related to it, particularly *Saccharomyces uvarum*. Ethanol is produced by the Embden–Meyerhof–Parnas (EMP) pathway. Pyruvate, produced during glycolysis is converted to acetaldehyde and ethanol. The overall effect is summarized as follows

$$Glucose + 2ADP \rightarrow 2\ Ethanol + 2CO_2 + 2ATP$$

Theoretical yields from 1 g glucose are 0.51 g ethanol and 0.49 g $CO_2$. However, in practice, approximately 10% of the glucose is converted to biomass and yields of ethanol and $CO_2$ may reach 90% of theoretical values. The ATP formed is used to supply other cell energy requirements.

The yeast-cell envelope contains a plasma membrane, a periplasmic space and a cell wall consisting mainly of polysaccharides with a small amount of peptide material. The wall component is a semi-rigid but solute permeable structure which provides considerable compressional and tensile strength to yeasts. Carboxyl groups of cell-wall peptides confer important flocculating characteristics on brewing yeasts, facilitating post-fermentation solid–liquid separation. Flocculation is thought to result from salt-bridge formation between calcium ions and these cell-wall carboxyl groups.

*Fermentation conditions*

Fermenting *S. cerevisiae* brewing yeast cells utilize the sugars, sucrose, fructose, maltose and maltotriose in that distinct order. Sucrose is first hydrolysed by invertase located in the extracellular periplasmic space. Sugars are transported across the cell membrane by either active or passive transport, mediated by inducible or constitutively-produced permeases. Maltose and maltotriose are hydrolysed intracellularly by α-glucosidase. *Saccharomyces uvarum* (*S. carlsbergensis*) is taxonomically distinguished from *S. cerevisiae* in that it can also utilize melibiose. *Kluyveromyces fragilis* and *Kluyveromyces lactis*, which unlike *S. cerevisiae* can ferment lactose, contain a lactose permease system for lactose transport into the cell where it is hydrolysed to glucose and galactose which enter glycolysis. With the exception of *Saccharomyces diastaticus*, which is not suitable for brewing, *Saccharomyces* yeasts are incapable of hydrolysing starches and dextrins. Use of starch-based materials for alcoholic fermentations requires addition of exogenous enzymes such as α- and β-amylases of malt or microbial enzymes such as α-amylase, amyloglucosidase (glucoamylase) and pullulanase. The major sugars of grape juice are glucose and fructose and since *S. cerevisiae* preferentially metabolizes the glucose, any unfermented sugar left in the resulting wine is fructose. In contrast, Sauterne wine yeasts ferment fructose faster than glucose.

Brewer's wort, produced from barley malt contains 19 amino acids and a range of other nutrients. These amino acids are assimilated at different rates during fermentation. A general amino acid permease (GAP) can transport all basic and neutral amino acids except proline. At least 11 other more specific amino acid transport systems exist in yeast. Proline permease is repressed by other amino acids and ammonia.

While alcoholic fermentations are largely anaerobic, some oxygen is needed to enable the yeast to synthesize some sterols and unsaturated fatty acid membrane components. Brewer's wort normally contains sub-optimal levels of sterols and unsaturated fatty acids, but if the medium is supplemented with oleic or oleanoic acid, the requirement for oxygen disappears.

Many strains of *S. cerevisiae* can attain ethanol concentrations of 12–14%. Interest has developed in the use of high alcohol tolerant yeasts in high-gravity brewing processes and in alcohol production for distillation with a view to increasing plant productivity and decreasing distillation costs. Selected strains are able to produce 18–20% alcohol although fermentation rate is generally greatly reduced as ethanol concentration increases. Grape juices with very high sugar concentrations are only fermented by osmophilic yeasts such as *Saccharomyces rouxii* and *Saccharomyces bailli* which have a high capacity to ferment fructose. Plasma membrane phospholipid composition is important for ethanol tolerance. Increased ethanol tolerance is observed when membrane unsaturated fatty acid content is increased. Alcohol tolerance may be enhanced by supplementing the growth medium with unsaturated fatty acids, vitamins and proteins. Physiological factors such as the method of substrate feeding, intracellular ethanol accumulation, osmotic pressure, and temperature all contribute to yeast ethanol tolerance. Yeast glycolytic enzymes hexokinase, glyceraldehyde-3-phosphate dehydrogenase and pyruvate decarboxylase have been shown to be sensitive to ethanol concentration.

For yeasts, pH values between 3 and 6 are most favourable for growth and fermentation activity. Fermentation activity is higher at higher pH values and there is a noticeable lag in fermentation activity at pH 3–4. pH affects formation of by-products. High pH values increase glycerol formation. The pH of grape must is usually in the range 3.0–3.9 because of its high acid content (5–15 g/l), mainly tartaric and malic acids. Since most bacteria, with the exception of acetic acid and lactic acid bacteria, prefer more neutral pH values the susceptibility of wine musts to infection is greatly reduced.

Temperature optima are distinctly different for yeast fermentation, yeast respiration and cell growth. Fermentation rate generally increases with temperature in the range 15–35°C and glycerol, acetone, butane-2,3-diol, acetaldehyde, pyruvate and 2-ketoglutarate levels increase in the fermentation broth. Formation of higher alcohols is also temperature-dependent. With white wines, lower fermentation temperatures produce fresher and more fruity wines, and the risk of bacterial infection and resultant volatile acid production is reduced. Higher temperatures of 22–30°C are used for production of red wines, fermented on the skins, leading to increased colour extraction and production of a rich aroma.

*Organoleptic compounds*

For production of alcoholic beverages, fermentations must be controlled so that, on the one hand, carbohydrates and other nutrients are assimilated for conversion to alcohol and desirable characteristic flavour components are produced, and on the other hand, undesirable flavour component formation is minimized. Flavour components include other alcohols, esters, carbonyl compounds, organic acids, sulphur compounds, amines and phenols.

Quantitatively, and in terms of their flavour-producing effects, the most important compounds present in beverages and spirits are the higher alcohols, also referred to as fusel alcohols or fusel oils. The most significant higher alcohols in beers and wines are amyl alcohol, isoamyl alcohol and 2-phenethanol. Red wines have higher concentrations of these compounds than white wines. Distilled beverages have a rather different spectrum of higher alcohols which includes butanols and pentanols. The polyhydric alcohol, glycerol, is present in almost all beverages and spirits. In wines, glycerol at concentrations of up to 1% (w/v) confers body to the drink.

In many beverages and potable spirits, esters constitute an important group of volatiles due to their penetrating fruity flavour. Of these, ethyl acetate is present in organoleptically important concentrations in many drinks. Other important esters include ethyl formate and isoamyl acetate.

Acetaldehyde, which has undesirable organoleptic properties, is produced as an intermediate during alcoholic fermentations. High levels of acetaldehyde can be caused by high yeast pitching rates or high aeration. Diacetyl and pentane-2,3-dione, which are formed outside the yeast cell by oxidative decarboxylation of $\alpha$-acetolactate and $\alpha$-acetohydroxybutyrate, have characteristic off-flavours and aromas. Diacetyl and pentane-2,3-dione may be reduced by the yeast and the presence of excess diacetyl results in beer when $\alpha$-acetolactate occurs at a stage when the yeast cells have settled out or have lost their ability to reduce diacetyl to acetoin. Excess diacetyl can also occur in beer due to the presence of beer-spoilage organisms such as *Pediococcus* and *Lactobacillus*.

During the primary fermentation of beer worts, considerable amounts of $H_2S$ gas are produced by reduction of sulphur compounds. While small amounts of sulphur compounds may be acceptable in beer, and in normal beer $SO_2$ is usually present at concentrations below its taste threshold, in excess, they produce unpleasant off-flavours. Sulphur dioxide has a positive effect in alcoholic fermentations in that it binds acetaldehyde and also inhibits some of the undesirable oxidative reactions. In wine musts, $SO_2$ inhibits undesirable micro-organisms including acetic acid and lactic acid bacteria as well as some naturally-occurring yeasts which produce excess volatile acids, pyruvate and $\alpha$-ketoglutarate. Apart from the unpalatable taste of acetic acid, it also inhibits yeast fermentations particularly in conjunction with ethanol. *Saccharomyces cerevisiae* is more susceptible to this inhibition than *Saccharomycoides ludwigii* and *Schizosaccharomyces pombe*. For musts which are low in total acidity, it is particularly desirable to use $SO_2$ to inhibit the malo–lactic fermentation by lactic acid bacteria, preventing further diminution of the acid level. Increased $SO_2$ concentration can delay the onset of fermentation.

## MALO–LACTIC FERMENTATION

The low pH of must and wine provides selective medium conditions for growth of lactic acid and acetic acid bacteria, although survival of the acetic acid bacteria is usually short lived due to the reducing properties of the medium. The lactic acid bacteria of wine are facultative anaerobes or microphilic organisms which are homofermentative (all *Pediococcus* and some *Lactobacillus* species, producing lactic acid from glucose) and heterofermentative (all *Leuconostoc* and some *Lactobacillus* species, producing lactic acid, ethanol and $CO_2$ from glucose). Malic and tartaric acids which occur in wine in higher concentrations as well as citric acid which occurs in lesser amounts, can be metabolized by lactic acid bacteria. In northern wine-making regions, bacterial reduction of acids may be desirable whereas in other regions, where acid concentration is low, it is important to prevent further loss of acidity. Chemical methods for acid reduction also exist. Reduction of acids can prevent spoilage of wines of high pH. At pH 3.3–3.4 *Leuconostoc oenus* ferments malic acid but *Pediococcus* species do not. Species of *Leuconostoc* are well adapted to low pH and are generally preferred whenever a malo–lactic fermentation is desired. A *Pediococcus* malo–lactic fermentation is often accompanied by formation of undesirable diacetyl and histamine.

BREWING

### Raw materials and malting

Traditionally-malted cereals have been used in brewing because the malting process results in the production of the requisite enzymes for extraction and conversion of the starch to fermentable sugars. In the malting process, grains are steeped at about 10–25°C for a period of 48–60 hours during which time grain moisture content is raised from 10–12% to 44–50%. After steeping, the moist grain is allowed to germinate at temperatures from 15–21°C, depending on malt type. Germination is carried out in drums or in perforated floor compartments with controlled humidity and air flow. Various hydrolytic enzymes including α-amylase, β-glucanases and peptidases, are produced and enzymatic degradation of hemicellulose, β-glucan, peptides and some starch occurs. β-Amylase, which is present in ungerminated barley bound to proteins, is released during this process. After the desired degree of modification the malt is kilned. This involves drying the malt to a moisture content of about 6% using a temperature of 65°C and then at 80–85°C, reducing the moisture level to about 4.5%. This process stops the germination respiratory processes, inhibits microbial development and promotes the Maillard reaction, thereby contributing a flavour and darkened colour to the malt.

Good quality malt has sufficient enzyme activity to facilitate extraction and conversion of mashes containing up to 60–70% of unmalted cereals in addition to malt. Industrial microbial enzymes can replace malt as a source of enzymes for wort production. Consequently, cereal-based alcoholic beverages can now be produced from a wide variety of individual cereals including barley, wheat, rice and corn, and cereal combinations, with or without the inclusion of malted cereal.

### Mashing and filtration

During mashing, the raw materials are extracted with water, and carbohydrate and other nutrients required for fermentation are solubilized, hydrolysed enzymatically and extracted from the cereal under controlled heating conditions.

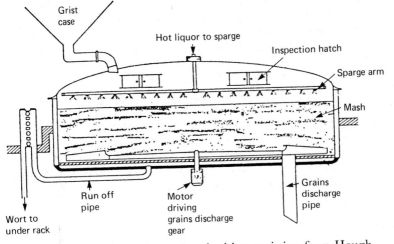

**Fig. 7.1** Infusion mash tun (reproduced with permission from Hough, 1985).

The water:grain ratio in mashing is usually 2.5–3.0 to 1.0. The temperature-mashing programme varies from one brewery or distillery to the next and will depend on considerations such as the nature of the cereals being used, the temperature-related properties of the enzymes, the desired specification of the product and the nature and capacity of the mashing equipment. Most mashing systems in current use are classified as infusion or decoction systems or a combination of both. In infusion mashing, using well-modified malt, the grain is fed into a mash tun, containing a filter-plate bottom (Fig. 7.1), partially filled with hot water. The temperature is held constant at about 65°C to allow enzyme conversion and wort extraction. The wort is then run off through the filter and the mash is sparged to wash out residual sugars until the specific gravity of the run-off drops to about 1.005.

In decoction systems, the grist is mashed in at a lower temperature, about 50°C, and the temperature is raised by transferring a portion of the mash to a 'kettle', boiling it and returning it back into the remainder of the mash. In a double-decoction mash, this operation is carried out twice (Fig. 7.2a). Following mashing, the wort is separated from the spent mash by filtration in a vessel called a lauter tun (somewhat similar to a mash tun but having a smaller mash depth), or by use of a mash filter (plate and frame).

The optimum temperature of malt protease, $\alpha$-amylase, $\beta$-amylase and $\beta$-glucanase dictate mashing conditions. Infusion mash temperature conditions favour amylase and $\beta$-glucanase activity and are most satisfactory with well-modified malts where considerable protein degradation has already occurred during malting. Decoction mashes usually have temperature stands at 50–55°C for optimal proteolytic activity, at 63–65°C for optimal degradation of starch and $\beta$-glucans and at 73°C for optimal wort separation. In North America, maize and rice grits, which require cooking for good starch gelatinization, are widely used as

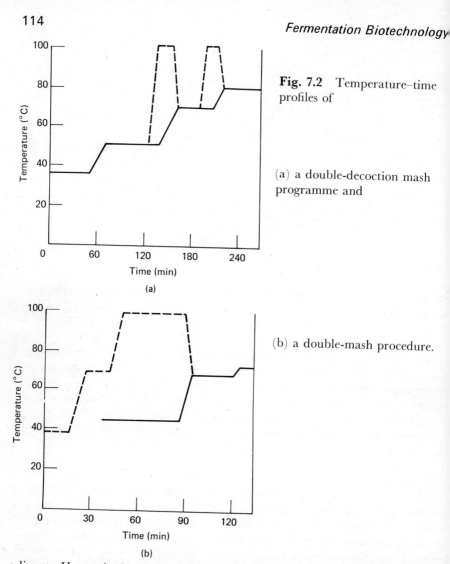

**Fig. 7.2** Temperature–time profiles of

(a) a double-decoction mash programme and

(b) a double-mash procedure.

adjuncts. Here a double mash system is used (Fig. 7.2b). The temperature of the first mash, containing the grits and some malt, is raised to 65–70°C, to reduce the viscosity of the starch prior to boiling. During this process, the main malt mash is held at a temperature of 45–50°C to promote proteolysis. By combining the contents of the grits mash with the main mash, the temperature is raised to 67°C which promotes amylolysis. The mash is then heated to 73°C to reduce viscosity before filtration. Where a mash is predominantly converted using microbial enzymes, the temperature stands will reflect the temperature activity and stability properties of these enzymes.

*Wort separation*

Following wort clarification, the wort is heated in the brew kettle and boiled for

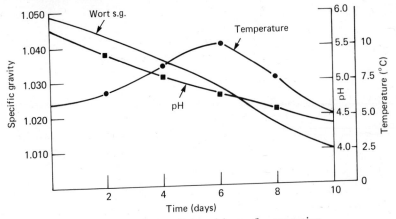

**Fig. 7.3** The time course of a traditional lager fermentation.

60–90 min. Wort boiling promotes (a) enzyme inactivation, (b) wort sterilization, (c) precipitation of proteins and tannins, (d) extraction of bitters and other substances from hops, (e) colour and flavour production from caramelization of sugars, melanoidin formation and tannin oxidation and (f) distillation of volatile material. The protein precipitate formed during boiling as well as the spent hops and other solids are separated, sometimes with the aid of flocculating agents, in a whirlpool separator or by a combination of a settling tank and filter or centrifuge. After separation of the solids or trub, the wort is cooled in a plate heat exchanger to the starting fermentation temperature and aerated to a dissolved oxygen concentration of 5–15 mg/l to promote yeast unsaturated fatty acid and sterol synthesis.

*Beer fermentation*
The wort is inoculated with yeast to give a count of about $10^7$ cells per ml wort or greater if a faster fermentation rate is required. Traditionally, top-fermenting yeasts, which ferment at 15–22°C and tend to rise to the surface towards the end of fermentation and can therefore be skimmed off, were used in ale production. Yeasts, which ferment in the range 8–15°C and which sink to the bottom towards the end of fermentation (bottom-fermenting yeasts), were generally used in lager production. However, particularly with the introduction of large cylindro-conical shaped fermenters and centrifuges to brewing operations, the distinction between top- and bottom-fermenting yeasts has tended to disappear.

In typical lager fermentations the starting yeast count is $10^7$ cells/ml, initial carbohydrate concentration is 12°P (specific gravity = 1.050) and initial temperature is about 10°C. After about three days, the yeast population has increased 4–5 fold. Temperature tends to rise as the fermentation proceeds and may require cooling when the desired maximum is reached. Carbohydrate concentration drops to 2–2.5°P (specific gravity = 1.008–1.010) after about five days. When the temperature is maintained between about 6 to 10°C, fermentation takes about 10 days. The course of a traditional lager fermentation operated at the lower temperature range is illustrated in Fig. 7.3. All fermentations conducted at temperatures between 15 and 22°C are usually completed in about three days.

During the fermentation process, pH drops about 1 pH unit from an initial value of approximately 5.2. Many acids, most notably acetic acid, formed by oxidation of acetaldehyde, contribute to this pH drop.

Fermentable sugars which normally constitute 70–80% of the wort carbohydrates are converted to ethanol during fermentation. The residual 20–30% carbohydrate consists mainly of higher dextrins or limit dextrins which are not susceptible to attack by malt amylases because of the presence of $\alpha$-1,6-glycosidic linkages. In the production of low carbohydrate beer, these limit dextrins may be hydrolysed to fermentable sugar by microbial amyloglucosidase. Ideally, this enzyme should be incubated with wort prior to wort boiling so that it is inactivated by the boiling process. While amyloglucosidase is also effective in hydrolysing limit dextrins during the fermentation stage, the enzyme is not inactivated by subsequent pasteurization and may remain in the final product.

*Beer maturation and conditioning (lagering)*

Freshly-fermented beer requires a variety of treatments before it is finally packaged for distribution. Maturation involves a secondary fermentation by the residual yeast which is transferred in the beer from the primary fermenter. During this process, beer diacetyl and small amounts of residual maltotriose are assimilated and concentrations of some esters increase. Carbon dioxide, produced from the secondary fermentation, or sparged carbon dioxide, helps purge the beer of oxygen, $H_2S$ and unwanted volatiles. Additives for haze clarification, flavour, aroma and colour adjustment, foam stability enhancement and microbiological stability are sometimes incorporated. The times and temperatures used vary from brewery to brewery. Usually low temperatures, 2–6°C, are used over a storage period ranging from 4 days to 4 weeks.

After lagering the beer contains microbial cells, proteinaceous precipitates and colloidal matter, which are separated out by a variety of processes including flocculation, centrifugation and filtration. Microbiological stability is achieved by sterile filtration and/or by pasteurization. The overall brewing process is summarized in Fig. 7.4.

## WHISKY PRODUCTION

Whisky is the potable spirit produced by distillation of the fermented aqueous extract of malted barley and other cereals. Properties of whiskies vary depending on the raw materials used and the nature of the fermentation and distillation processes. Scotch whisky is made from all-malt or from a blend of all-malt whisky with lighter Scotch grain whisky, made from a mixture of malted barley and unmalted, uncooked barley. A characteristic peat flavour of Scotch malt whisky derives from the use of peat to dry the malt grains. Bourbon whisky, which evolved in the United States, and Canadian Rye whisky are produced from mashes in which maize and rye predominate respectively, with malt and other cereals as more minor components.

*Food fermentations*

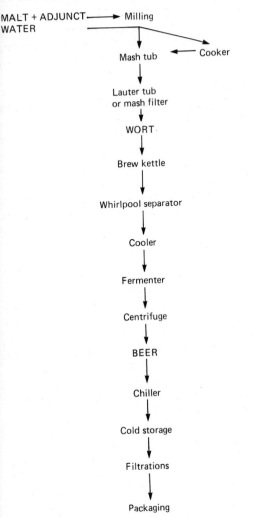

**Fig. 7.4** Summary of the brewing process.

*Mashing*
In Scotland and Ireland, clear wort is produced during mashing, using an infusion mash process. Wort preparation processes are similar to those used in brewing except that amyloglucosidase is added and remains active throughout the fermentation process to optimize fermentability of the wort. In North America, owing to their higher starch gelatinization points, unmalted cereals, particularly maize, must be pre-cooked before being combined with the other mash components for enzyme conversion and extraction. Batch and continuous pressure-cooking systems are used at temperatures ranging from 110 to 170°C.

In the American batch-cooking process, milled corn is slurried with water and fed into an agitated cooker at a concentration of about 25% w/v. Temperature-stable liquefying α-amylase partially hydrolyses the gelatinized starch (pre-liquefaction), reducing the mash viscosity sufficiently to allow agitation to continue. The enzyme is inactivated as the temperature increases above 110°C and the remaining starch is gelatinized. The mash is cooled to 70–100°C, depending on the temperature-stability characteristics of the α-amylase to be used in the post-liquefaction stage. When the starch is completely liquefied by α-amylase, the temperature is further reduced to 60°C and amyloglucosidase is added to convert the dextrins to glucose (saccharification). In continuous processes, for cooking and liquefaction of the corn, better process control and more efficient equipment usage is achieved. The mash temperature is raised to 150°C by steam injection in a jet cooker and then flash cooled to 80–90°C for a post-liquefaction stage lasting 30–60 min before being transferred to a saccharification tank. When the final mash contains a mixture of malt and corn, these may be combined for the saccharification stage, allowing the malt enzymes to contribute to these hydrolytic processes.

In whisky production, a proportion of the still bottoms (stillage) from previous whisky distillations may be mixed with water used to mash in the cereal grains. Stillage reduces the mash pH somewhat, provides buffering capacity, is a source of nitrogenous compounds for yeast nutrition and is an effective waste-recycling measure.

*Whisky fermentation*

Fermentations of Irish and Scottish whiskies are carried out on clear worts. The unseparated grain mash is used directly in North American whisky fermentations.

In contrast to brewery worts, whisky worts are not hopped. In addition, whisky worts are not boiled so that enzyme activity continues throughout the fermentation and more contaminating bacteria are present in the wort. The majority of aerobic bacteria die in the anaerobic fermentation environment. On the other hand, *Lactobacillus* and *Streptococcus* species can grow under these conditions. Provided their numbers are controlled, compounds excreted by them contribute to the organoleptic qualities of whisky. When present in excess, an overall decrease in spirit yield is observed and undesirable organoleptic properties may result. Distillery yeasts are selected on the basis of their suitability to the distillery environment and their ability to produce a combination of high ethanol yield and the desired flavour components present in distillates.

Where fermenters are equipped with efficient cooling systems, fermentation temperature may be set at 33°C and maintained there by cooling. Distilleries not equipped with efficient cooling systems initiate the fermentation process at lower temperatures, around 20°C, such that the temperature does not rise above 33°C during the fermentation. Lower temperatures are often considered desirable during the initial fermentation stage, when the yeast is growing, as the higher temperatures retard yeast growth and encourage bacterial growth. A typical malt whisky fermentation produces an ethanol concentration of 9–12% in 36–48 h from a wort having an initial gravity of 1.050–1.060.

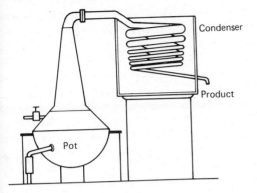

Condenser

Product

Pot

**Fig. 7.5** A pot still.

*Whisky distillation and maturation*

Whisky stills are designed to selectively remove volatile compounds, particularly the flavour-producing compounds, and to create additional flavour-generating compounds as a result of the chemical reactions which take place during this process. The major components of the traditional pot still (Fig. 7.5) are the copper pot and 'worm' condenser. These stills, which provide little rectification, are used exclusively to distil Scotch malt whisky and most Irish whisky and produce products with high quantities of esters, aldehydes and higher alcohols (congeners). Continuous stills are generally used for processing grain whisky. Two columns are employed, one called the analyser, the other the rectifier. The feed is pre-heated, then vaporized in the analyser and the majority of the water and congeners is removed in the rectifier. The extent of congener removal will depend on the product. Some or all of the congeners may be desirable in the whisky whereas in neutral spirit congeners are undesirable.

Whisky is stored normally in oak casks to mature, at least for the periods of time required by law, or longer to produce a product of particular quality. During maturation, a combination of physical and chemical reactions take place, which include extraction of components from the wood, oxidation of distillate congeners and wood extracts, and interaction of congeners and extracted wood chemicals. The final process stage involves blending of matured whiskies to produce a variety of individual products.

## WINE-MAKING

The process for producing white wine involves crushing of the grapes and separation of the seeds and skins from juice using screw or hydraulic presses to produce free-run juice for fermentation. Sulphur dioxide is added to the juice to inhibit oxidative enzymes and growth of undesirable micro-organisms. The pressed juice is freed from particulate matter by centrifugation, filtration or sedimentation. The juice is inoculated with wine yeast at temperatures below 20°C. Temperature rises during the fermentation process and must not exceed

36–38°C which kills the yeast. At the end of fermentation, an ethanol concentration of 8–15% by volume is achieved. The wine is separated from yeast, clarified, stabilized and stored.

With red wine, the fermentation is carried out on the skin, for 3–10 days or longer, using crushed grapes as starting material. Fermentation temperature is maintained between 20–30°C, to prevent excessive tannin extraction and the ethanol being produced accelerates colour extraction. Colour can also be extracted by a short high-temperature treatment of the crushed grapes at 85°C and this usually results in less tannin being extracted. In the case of red wines, where a malo–lactic fermentation is desired, the separation of yeast from wine may be delayed, resulting in some amino acid release through autolysis, which encourages growth of bacteria responsible for the secondary fermentation.

Large-scale rack and cloth filters are used for initial wine clarification. The rough filtered wine is then fined to aid flocculation and absorption of suspended solids and again filtered using a plate and frame filter. The wine is then aged in tanks for periods ranging from a few weeks to up to two years prior to bottling. During this period some chemical reactions, including oxidations and esterifications, take place.

## Cheese-making

The manufacture of cultured dairy products is considered to rank second only to production of alcoholic beverages among the industries which rely on microbiological processes. While hard and soft cheeses represent perhaps the most important cultured dairy products, other significant product types include yoghurts, sour cream, buttermilk, acidophilus milk, ripened cream butter, kefir, koumiss and Bulgarian milk. This section is confined to a discussion of the cheese-making process.

The general stages of cheese-making are

(1) Milk pretreatment
(2) Coagulation
(3) Separation of solid curd from fluid whey
(4) Forming of curd
(5) Cheese ripening

Milk with a relatively low bacterial count is generally best for cheese-making as a high count can lead to problems associated with excessive gas production and off-flavour development. Several milk enzymes can exert an effect on cheese-making and cheese quality. Lactoperoxidase inhibits some lactic acid bacteria important in cheese manufacture. Lipases and proteases may modify the flavour and rheological properties of cheese by hydrolysis of milk fat and casein, respectively. Milk microbial and enzyme activity may be reduced by pasteurization. Cottage and cream cheeses usually employ pasteurized milk. Homogenization of milk, which reduces the size of milk fat globules, is used in the manufacture of cream cheese, some soft cheeses and cheese spreads. The resultant curd formed is weaker and lipolysis and flavour development is generally accelerated. Cheeses are made from milk with altered fat content. Cream, used in cream cheese manufacture, has a higher fat content than normal milk, whereas skim milk is used in cottage cheese. The natural cheese colour may be bleached using benzoyl peroxide, masked using artificial colouration, or darkened using natural colour extracts such as annatto.

Milk coagulation is generally achieved by the combined processes of milk pH reduction and milk-clotting enzymes. Milk acidification is achieved by use of lactic starter cultures to ferment lactose to lactic acid (see Chapter 6 on starter cultures for the dairy industry). Acid production reduces the solubility of casein as the pH approaches 4.6, the isoelectric pH of casein, and accelerates enzymatic cheese coagulation. Further acid-reduction aids whey syneresis (extraction) from curd and controls development of an undesirable microbial flora. The acids and other metabolites produced by lactic cultures contribute to cheese flavour. Cheese-clotting enzymes are characterized by their ability to specifically hydrolyse the kappa-casein fraction of milk without attacking the other major casein fractions. Kappa-casein in milk stabilizes the milk casein micelles in the presence of calcium in a colloidal suspension. Hydrolytic release of a macropeptide from kappa-casein, by the milk-clotting enzyme, destabilizes the micelle leading to coagulation.

Suitable cheese-clotting enzymes must have a high ratio of cheese clotting to proteolytic activity to avoid excess protein degradation which would reduce curd yield. The most widely used clotting enzymes are calf rennet, extracted from the fourth stomach of calves, 50:50 mixture of calf rennet and porcine pepsin and microbial rennet, produced by *Mucor miehei*. Milk coagulation can be accelerated by increasing the temperature to 45°C or delayed at low temperatures. The kappa-casein hydrolysis stage can be separated from the coagulation stage by cold-renneting and later warming the milk to achieve coagulation, a technique which may be used in continuous cheese-making.

Following coagulation, the solid curd and the liquid whey are separated. This process of whey syneresis is accelerated by various means including cutting of the curd, increasing temperatures, pH reduction by lactic bacteria and various physical separation techniques.

A variety of reactions, leading to flavour development, take place during cheese ripening. In some cheese, specific micro-organisms such as *Propionibacterium* (Swiss Cheese), *Penicillium roqueforti* (Blue), or *Penicillium camemberti* (Camembert), provide the dominant flavour. In other cheese enzymes, such as pregastric esterases (Romano and Provolone) make a major contribution to cheese flavour. Other cheeses, like Cheddar and Gouda, depend on the combination of complex ripening reactions for flavour production.

Clearly, the process for manufacture of cheeses varies from variety to variety. A flow diagram for production of cheese is provided in Fig. 7.6.

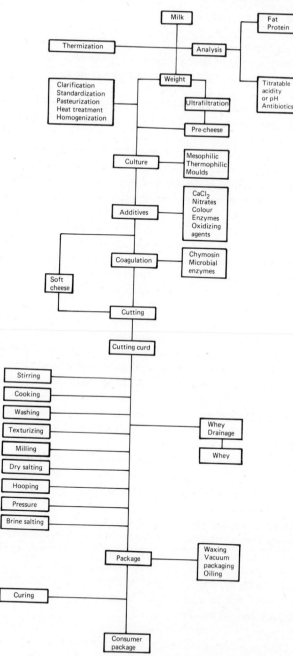

**Fig. 7.6** Flow diagram of a cheese-making process (reproduced with permission from Irving and Hill, 1985).

## Bread-making

Most bread is produced by mixing flour, especially wheat flour, with water, some yeast, salt, sucrose and shortening. Following several minutes of mixing the ingredients, the bread is incubated at a temperature of 25–35°C to promote the fermentation of dough sugars by *S. cerevisiae* to alcohol and carbon dioxide gas bubbles which are trapped in the dough. After fermentation, the process of baking the dough evaporates the ethanol but the gas bubbles remain, conferring texture on the bread. The overall result of the yeast fermentation is inflation of the dough, modification of the gluten structure, development of a light spongy texture and contribution of flavour compounds to the bread. Dough-fermentable sugars include the added sucrose and also glucose and maltose which may be produced from the starch by the action of cereal α- and β-amylases or added fungal α-amylase.

The main methods of making wheat bread dough are as follows:

(1) The straight dough process, where all the ingredients are at once mixed together and fermented as a batch
(2) The sponge and dough processes, where the fermentation is mainly carried out on a portion of the dough called the sponge
(3) Continuous mixing processes, followed by fermentation of either the bulk dough or loaf size dough portions
(4) The liquid ferment process, a variation of the sponge and dough method, particularly suitable for continuous dough-mixing processes, where the sponge stage is replaced by a liquid-pumpable fermentation system.

Non-fermentation methods of raising dough include mechanical whipping/aeration processes and use of sodium bicarbonate or ammonium bicarbonate as a source of gas bubbles.

Sour doughs are produced mainly from rye flour but also from wheat flour. Rye flour lacks elasticity and lactic acid bacteria acidify the rye dough, rendering it more suitable for baking, and a yeast fermentation raises the dough. Enzymes, from micro-organisms involved in the sour dough, degrade the cereal pentosans and proteins lowering rye dough viscosity. The lactic and acetic acids produced during the fermentation confer characteristic flavour properties on the bread.

## Fermented soya-based foods

Food fermentations, using soybeans and other raw materials have been practised for many centuries especially in the Orient. Among the most important fermentations are the processes for production of soy sauce, miso and tempeh. Outline processes for their manufacture are illustrated in Fig. 7.7.

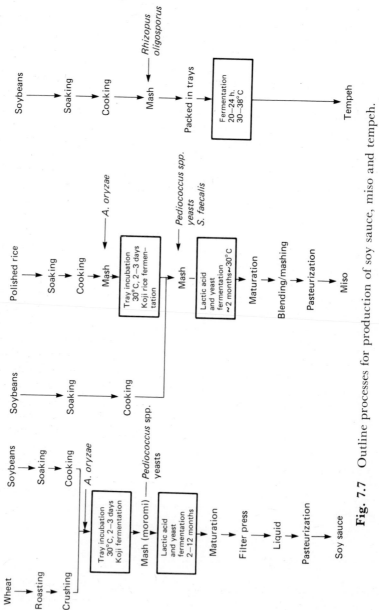

**Fig. 7.7** Outline processes for production of soy sauce, miso and tempeh.

While a variety of soy sauce types exist, the predominant sauce produced in Japan, called koikuchi, is produced by fermentation of a wheat soybean mixture, to obtain a seasoning having a strong aroma and dark reddish-brown colour. The process initially involves a primary fermentation of a 50:50 mixture of soaked, steamed soybean and roasted crushed wheat, having a moisture level of 27–37%, spread on trays, and incubated with strains of *Aspergillus oryzae*, having good proteolytic and amylolytic activity, for 2–3 days at 25–30°C. After addition of saline water, to raise the mash sodium content to about 18%, the mash or 'moromi' is transferred to large vats where a secondary fermentation takes place, involving halophilic bacteria and osmophilic yeasts. Process conditions include temperature maintenance at 35–40°C, intermittent stirring and a fermentation time of 2–12 months. At the end of the secondary fermentation, the liquid sauce is recovered and pasteurized. Enzymes produced in the primary or koji fermentation hydrolyse proteins to peptides and free amino acids and convert starch to simple sugars. These breakdown products are in turn metabolized by other microorganisms to produce a wide variety of flavour and aroma compounds. By use of pure culture fungal, bacterial and yeast inocula, better fermentation process control is obtained and product of consistent quality and flavour is routinely achieved.

In the Orient, fermented soybean pastes have traditionally been prepared in the home and used as a base for soups and sauces. In Japan, the product, which is called miso, is now manufactured commercially on a large scale. The process initially involves koji fermentation of soaked steamed rice (or barley or soybean which are less frequently used), spread on trays with selected strains of *A. oryzae* at 28–35°C for approximately two days. This koji material is then mixed in vessels with soaked, steamed soybeans in a ratio of 1:2, with sodium chloride added to give a concentration of 4–13%. The mixture is inoculated with bacteria and yeast and fermented for between 1 and 52 weeks. The product is then blended and pasteurized. Various types of miso can be produced with different degrees of sweetness, saltiness and colour by varying koji raw materials, salt content, sweeteners and fermentation duration.

Tempeh production involves a tray fermentation of soaked steamed soybean inoculated with *Rhizopus oligosporus*, and incubated at 30–38°C for about 24 hours. The product which is usually sliced, salted and deep-fat fried for consumption, has a short shelf life.

## Meat fermentations

The majority of fermented meat products consist of dry or semi-dry sausages. Products usually have a moisture to protein ratio of between 1.5 and 3 to 1. A microbial fermentation step at optimal temperatures generally precedes subsequent heating and/or drying.

Microbial fermentation, resulting in acid development, aids processing in a number of ways. Fermentation contributes to product stability and shelf life, facilitates subsequent moisture removal and generates unique flavours and aromas. It is generally recommended that the pH of the product should be less than 5.3, based on the assumption that such pH values control *Staphylococcus aureus*. Fermentation processes to reduce the pH below 5.3 should conform to specified temperature–time safety guidelines which range from 23.9°C–43°C for incubation periods of 80–18 h. Commercial inoculants, such as *Pediococcus* species, having good lactic acid producing ability, are frequently used. *Staphylococcus carnosus*, a harmless coagulase-negative species, is routinely used in production of dry sausage. Nitrite has a positive effect on meat preservation as it controls the development of *Clostridium botulinum*. *Micrococcus* species, having the ability to reduce nitrates to nitrites, in a controlled manner, contribute to the meat-curing reaction as well as to *C. botulinum* control. Starter cultures may also be used in bacon processing to dissipate any residual nitrite present, thereby lowering or eliminating carcinogenic nitrosamine formation during frying.

Through greater use of commercial inoculants, it is predicted that the meat fermentations will be characterized by a high level of process control, giving reproducible product quality and consistency while maintaining highest safety standards.

## Vinegar

Vinegar, an aqueous solution of acetic acid, is produced in fermentation by oxidation of a dilute ethanol solution. The metabolic process involves conversion by alcohol dehydrogenase of ethanol to acetaldehyde and conversion by acetaldehyde dehydrogenase of hydrated acetaldehyde to acetic acid

$$C_2H_5OH \rightarrow CH_3CHO + H_2O \rightleftharpoons CH_2CH(OH)_2 \rightarrow CH_3COOH + H_2O$$

The fermentation raw material for alcohol (white spirit) vinegar is usually diluted purified ethanol. Wine, cider, malt and rice vinegars are produced by alcoholic fermentation of grape juice, apple juice, barley malt or rice mashes, respectively. The spirit vinegar fermentation also requires a nitrogen source and an appropriate combination of minerals and the fermented mash vinegars often require nutrient supplementation. Modern vinegar fermentations are highly aerated, submerged processes. The Frings acetator (Fig. 7.8) which is widely used for commercial vinegar production is a baffled fermenter containing a bottom-driven hollow-body turbine which rotates at 1450–1750 rpm. The rotating action of the turbine design enables air to be sucked through the hollow rotor and distributed radially over the whole cross-section of the fermenter.

Typical commercial processes, involving production of 12–15% acetic acid, are carried out in a semi-continuous manner. Acetic acid and alcohol concentrations at the start of the cycle are 7–10% and 5%, respectively. The fermentation proceeds at 27–32°C until the alcohol concentration drops to 0.1–0.3%, at which point a quantity (about one-third) of the vinegar is discharged and the vessel is filled with new mash containing 0–2% acetic acid and 12–15% ethanol and the cycle is repeated. An alkalograph measures the ethanol concentration and enables the discharge and refilling operations to be carried out automatically. The

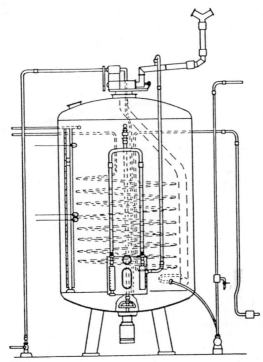

**Fig. 7.8** Diagram of a section of a Frings acetator (reproduced with permission from Greenshields, 1978).

discharge and filling operations must be carried out without interruption of the aeration/mixing process in a manner which avoids total ethanol depletion from the fermentation broth. The continuous rapid mixing is essential to minimize concentration gradients. Cycle times vary from 24–48 h.

Acetic acid bacteria, which oxidize ethanol to acetic acid and can exist at low pH values, come from the closely-related genera *Acetobacter* and *Gluconobacter*. Pure cultures of these organisms are characterized by their high degree of variability and in industrial fermentations mixed cultures will consequently develop from a pure culture. Industrial cultures are selected to tolerate high acidity and to yield high acetate production rates. These bacteria are extremely sensitive, are killed by lack of oxygen and lack of ethanol and are also damaged by acetate and ethanol concentration gradients. Sensitivity to lack of oxygen increases with increasing total concentration of acetic acid plus ethanol. Nevertheless, with efficient aeration, an oxygen utilization of 80% can be achieved without adverse effects on the fermentation. Over-oxidation, that is conversion of acetic acid to $CO_2$ and $H_2O$, can be avoided by maintaining acetic acid concentrations above 6% and avoiding total depletion of ethanol.

# Chapter 8

# *Industrial chemicals*

### Bulk organic chemicals

Nearly all commodity chemicals are currently produced from petroleum and natural gas resources. Problems associated with this reliance on petroleum include fluctuating petroleum costs, uncertainty of supply and ultimately concerns relating to the depletion of these non-renewable fossil fuels. It is widely predicted that, as world petroleum stocks become depleted and prices rise, petroleum companies will switch to coal as the major source for chemical feedstocks and energy production. Most petroleum feedstock-processing schemes can readily be adjusted to accommodate coal as a raw material and consequently major new investment in chemical plants is not required. However, because coal, like petroleum, is a non-renewable resource, a gradual transition to the use of biomass feedstocks is expected to provide solutions to the long-term problems associated with petroleum and coal depletion.

Currently, ethanol is being produced by fermentation, as a fuel and commodity chemical, in the United States and Brazil, where corn starch and cane-sugar are used, respectively, as major raw materials. While there is scope for expansion of processes for use of sugar- and starch-based feedstocks, their widescale use for chemical and energy production would place strains on world food supplies. Extensive use of biomass for energy and commodity chemicals' production would most likely rely on lignocellulose as raw material. Biomass feedstocks have an inherent disadvantage in that their energy yield per unit weight is approximately one-third that of petroleum. Consequently, costs in transportation of biomass materials to chemical processing plants would be significant.

Raw materials for microbial conversion into useful chemicals must be pre-treated by various methods prior to fermentation. Cane-juice sugar is extracted

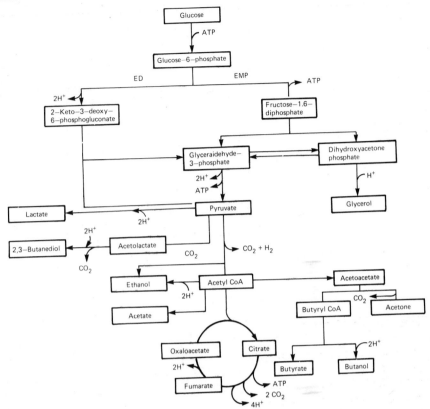

**Fig. 8.1** Metabolic pathways for biosynthesis of various chemicals (reproduced with permission from Ng *et al.*, 1983).

by milling and washing procedures. Preparation of cereals for fermentation involves milling, mashing, heating and conversion of starch to glucose by enzymatic liquefaction and saccharification. Lignocellulose biomass requires more severe pre-treatment and hydrolysis procedures. Current pre-treatment methods, to remove lignin and expose the crystalline cellulose, including use of acids, bases, steam explosion and mechanical procedures, add significantly to processing costs. Bio-delignification processes, using micro-organisms such as *Chrysosporium pruinosum*, have potential for lignin removal, but current microbial reaction rates are much too slow to be economically feasible. In addition, the crystalline structure of cellulose makes it more intractable to acid and enzymatic hydrolysis than polysaccharides like starch and the development of improved biotechnological processes for production and application of cellulases is considered to be a prerequisite to the efficient utilization of lignocellulose biomass resources.

The metabolic pathways for production of various chemicals by fermentation are illustrated in Fig. 8.1. Current commercially-viable fermentations include processes for production of ethanol, citric acid, gluconic acid, itaconic acid and lactic acid. Industrial fermentations have been used in the past for production of acetone, butanol, glycerol, and fumaric acid but less-expensive petroleum-based chemical routes are currently used for their manufacture. Research is being carried out to further develop the acetone–butanol fermentation in order to make it commercially viable. Other fermentation processes for production of acetic acid and 2,3-butanediol as chemical feedstocks have real commercial potential.

ETHANOL PRODUCTION

The use of fermentation processes for production of industrial alcohol has several advantages

- Fermentation technology currently exists
- Alcohols can be produced from renewable resources
- Wastes and inedible agricultural products may be used as substrates
- Recovery processes are relatively simple
- Alcohol fuels burn more cleanly than gasoline fuels

Industrial alcohol is produced in various grades. The majority is 190 proof (92.4 wt%) alcohol, used for chemical, cosmetic and pharmaceutical applications. Anhydrous ethanol (99.8 wt%) is required for specialized chemical applications. For fuel used in gasohol (see below), nearly anhydrous (99.2 wt%) alcohol is used. Distillation is used almost exclusively as a means of ethanol recovery. Simple one- or two-stage columns with a stripping and rectifying section are used to produce alcohol up to a concentration of 95.7 wt%, where ethanol and water form an azeotrope. The volatilities of water and ethanol are then identical. Production of anhydrous alcohol generally requires addition of a third chemical such as benzene to alter the distillation equilibrium.

The major carbohydrate sources for ethanol fermentations are sugars from beet and cane, starch from cereals and root crops, cellulosic materials from wood and wastes and by-products of other processes such as sulphite waste liquor and whey. Currently, 95% of world fermentation alcohol is produced from hexoses using *S. cerevisiae*. Biological aspects of alcohol production by *S. cerevisiae* have been discussed in Chapter 7. Due to interest in utilization of hemicellulose, possible fermentation processes for production of ethanol from pentoses are also being investigated. While *Saccharomyces* strains do not metabolize xylose, a number of other yeasts can convert xylose to ethanol and lactate. *Clostridium thermosaccharolyticum*, *Thermoanaerobacter ethanolicus* and other thermophilic bacteria are also being investigated for use as pentose fermenters. While the organisms so far studied produce excessive quantities of undesirable side-products or only produce low alcohol concentrations, their capacity to ferment at higher temperatures might allow for continuous alcohol recovery, thus reducing end-product inhibition effects. The bacterium *Zymomonas mobilis* is also of interest because it converts glucose to ethanol at a yield 5–10% higher than most yeasts. Its disadvantages are its low alcohol tolerance and its small size which makes cell separation difficult. There is also interest in using thermotolerant *Clostridium* species, including *Clostridium thermocellum* and *Clostridium thermohydrosulfuricum*, to convert cellulose to ethanol. Physical, chemical and enzymatic methods for conversion of cellulosic and hemicellulosic materials to fermentable sugars are also being investigated.

The technology for use of alcohol as a fuel dates back to the 1920s when Henry Ford's Model T was designed to run on alcohol, gasoline or any mixture of the two. However, alcohol was not used to any extent in motor fuel until the 1970s when oil prices increased dramatically. The sudden oil-price increases, together with the limited availability of fossil fuel resources, stimulated the search for new energy sources. Projects were initiated on the evaluation and development of a variety of technologies which might harness other kinds of energy, such as solar, wind, geothermic or tidal energy. The use of solar energy to transform $CO_2$ into biomass by photosynthesis and the use of micro-organisms to convert biomass sugars to ethyl alcohol thus found immediate application.

In the USA, a blend of 10% alcohol in gasoline, called gasohol, was introduced, and gasohol production was encouraged by the federal government by reducing the excise tax on gasohol. The USA government target was to achieve a fermentation production capacity of 1.8 billion gallons per annum by the mid-1980s. Brazil, the world leader in fermentation alcohol production, was producing 1.4 billion gallons in 1981–2 and set a target of 3 billion gallons per annum for 1987.

*The Gasohol Program*

By 1986, production capacity in the USA had reached 760 million gal/annum with the expectation that the original target of 1.8 billion gal/annum would be reached by 1990. Because of the reduced environmental polluting effect of ethanol, there is strong pressure in the USA to have all automobiles use gasohol. This would require an annual fermentation ethanol capacity of 8 billion gal/annum. There has also been some expansion of fuel alcohol production capacity in Canada.

The major feedstock for the production of fuel alcohol in the USA is corn. Both batch and continuous fermentation processes are used and both process types use continuous corn cooking procedures. A typical batch-fermentation process is outlined in Table 8.1. The yeast population is usually about $2 \times 10^8$ per ml and

**Table 8.1**  Typical batch-fermentation process for ethanol production from corn

(a) Corn milling to a screen size of $\frac{1}{8}$–$\frac{3}{16}$ inch
(b) Addition of water at a water:corn ratio of 3–4:1 by weight
(c) Pre-liquefaction of the mash using temperature-stable α-amylasse from *Bacillus licheniformis* at 60–66°C for 20–30 min
(d) Complete gelatinization of the corn starch by raising the temperature to greater than 150°C for 5–10 min
(e) Post-liquefaction of the starch, using a second dose of temperature-stable α-amylase, after reducing the mash temperature to 85°C
(f) Mash cooling to 32°C and transfer to fermenter
(g) Addition of amyloglucosidase and yeast to the mash initiating simultaneous starch saccharification and alcohol production in the fermenter
(h) Recovery of alcohol after a fermentation time of approximately 48 h

fermentation time is about 48 h. A few major USA alcohol producers use continuous fermentation processes. The corn is wet milled to produce a mash free of solids. By use of flocculant yeasts and a yeast recycle step, yeast populations of $6-8 \times 10^8$ per ml are achieved and the fermentation may be completed in 10–18 h.

## *The Brazilian National Alcohol Program*

The Brazilian National Alcohol Program was created in 1975 to reduce petroleum imports and to partially or totally substitute gasoline with anhydrous ethanol. The programme gave incentives to private enterprise to increase agricultural production of energy crops and their transformation into substitute petroleum derivatives and raw materials for the chemical sector. Ethanol was produced exclusively by fermentation of cane-sugar. The Government Aeronautical Technical Centre designed the Otto cycle engine to run on ethanol and transferred the technology to the motor industry. Financial incentives were provided to encourage motorists to use alcohol exclusively as a motor fuel. These policies resulted in a dramatic increase in use of both straight ethanol and also gasohol as motor fuels. The reduced utilization of crude oil for petroleum refining has resulted in a shortage of diesel oil and a large-scale diesel oil substitution programme is required, possibly using ethanol.

Parts of the programme are concerned with improving the efficiency of sugar-cane production, the alcohol fermentation stage and waste recycling. Strategies adopted were as follows:

(i) *Efficiency of sugar-cane production.* Problems of low sugar-cane productivity were to be addressed by use of better varieties of sugar-cane species, improvement of agricultural practice and control of disease and insect damage. It was assumed that with good agricultural practice, yields of cane of 76 ton/ha could be obtained and with three harvests being recovered in four years, an average productivity of 55 ton/ha/year could be obtained. Use of new plant varieties was expected to raise annual average productivity to 80 ton/ha. The achievement of four harvests in five years would raise the yield to 101 ton/ha. It was also hoped to increase the amount of fermentable sugar contained in the harvested cane over a period of about fifteen years from 13.5% to 17%. By improving mill design, it was considered feasible to increase sugar extraction efficiency from 91% to 97%.

(ii) *Alcohol production.* The strategy for improving the efficiency of the fermentation and distillation operation, resulting in decreased plant and labour investment, was to extend the period of operation of these processes beyond the 150–180 day harvesting and milling season to 200–300 days, by use of new improved methods of molasses and cane-juice storage. By improved control of temperature, pH and infection prevention, it was estimated that fermentation efficiency could be increased from 85% to 91%.

(iii) *Waste treatment.* The most important industrial wastes from cane alcohol production are stillage and bagasse. The stillage is used in Brazil mainly as a dried soil fertilizer, as a cattle feed supplement (after potassium reduction to prevent gastrointestinal disorders) and as a feedstock for methane production. Bagasse is used as a steam-generating fuel.

In Brazil, about half of the alcohol plants process cane directly as a raw material. Sugar-cane juice is recovered by a milling process which involves cutting, milling and rolling to capture the cells and express the juice containing 12–16% sucrose. The sugar solution, supplemented with any required nutrients, is inoculated with yeast and maximum ethanol productivity is achieved in 14–20 h in a batch process. Several fermenters are usually operated on a staggered basis to provide a continuous feed to the distillation plant. Most Brazilian distilleries use the 'Melle Boinot' process, which involves centrifugal recovery of the live yeast (usually 10–15% by volume of the total beer) and its re-inoculation into other fermenters.

## CITRIC ACID PRODUCTION

Citric acid, which is widely present in nature, has long been used in soft drinks manufacture as an acidulant, as an aid to jam setting and as a general additive in the confectionery industry. It was first commercially extracted from lemon juice, later synthesized from glycerol and other chemicals and finally first produced by industrial fermentation in 1923. By 1933, annual world production exceeded 10 000 t with greater than 80% being produced by fermentation. With the substitution of polyphosphates by sodium citrate in detergents around 1970, the annual market grew rapidly and is now estimated to be more than 300 000 t, exclusively produced by fermentation. Initially, surface culture methods were used employing *Aspergillus niger*. Submerged *A. niger* culture processes were introduced after the Second World War, and around 1977, a submerged process, involving *Candida* yeast, was commercialized.

The main features of citric acid industrial processes are summarized in Table 8.2. Surface culture using *A. niger* is still extensively used. While it is more labour intensive than submerged culture, power requirements are less. With submerged processes, well sparged air-lift fermenters are preferred which allow larger size vessels to be used for this high bulk fermentation product. Molasses is the most widely used fermentation raw material for citrate production and variability of this material remains a major problem. For submerged *A. niger* fermentations, increased sugar concentrations stimulate citrate production and poor yields of citrate are obtained if the sugar concentration is less than $140 \, \text{kg m}^{-3}$. Nitrogen is usually supplied at a concentration of $0.1–0.4 \, \text{g l}^{-1}$. Addition of $NH_4^+$ during the fermentation increases citrate production.

Citric acid production by *A. niger* is extremely sensitive to $Mn^{2+}$ concentration. Production of the acid is drastically reduced at a manganese level as low as $3 \, \text{mg l}^{-1}$. It is therefore necessary to pre-treat molasses media with agents which complex with or precipitate manganese, such as hexacyanoferrate (HCF), or with copper which counteracts the effect of manganese by inhibiting its cellular uptake. These manganese-deficient conditions also favour production of small mycelial pellets with hard smooth surfaces, characteristically found in good citrate fungal fermentations. When manganese levels increase, fungal morphology becomes filamentous thereby dramatically increasing the viscosity of the broth with a concomitant rapid decrease in dissolved oxygen tension. Since citric acid

**Table 8.2** Summary of industrial citric acid fermentations

| Parameter | *Aspergillus niger* | *Aspergillus niger* | *Candida guilliermondii* |
|---|---|---|---|
| Fermentation type | Surface culture, depth 0.05–0.2 m | Submerged culture 40–200 m$^3$ stirred tank reactor, 200–900 m$^3$ air-lift reactor | Submerged culture 40–200 m$^3$ stirred-tank reactor, 200–900 m$^3$ air-lift reactor |
| Production fermenter inoculum | Conidia/spores ca. 150 mg or $2 \times 10^9$ spores/m$^3$ | Vegetative inoculum prepared in seed fermenter or direct spore inoculum | Inoculum prepared in seed fermenter |
| Fermentation pH | Initially 5.0–7.0 for *A. niger* germination/growth. Drops below 2.0 for citrate production phase | | pH 4.5–6.5 for growth. Can be allowed to fall to about 3.5 for citrate production |
| Temperature | 30°C | 30°C | 25–37°C |
| Aeration (function)* | – (oxygen transfer, cooling) | 0.5–1 vvm (oxygen transfer, mixing in air-lift reactor). High O$_2$ tension >140 m bar. Fermentation very sensitive to oxygen | 0.5–1 vvm (oxygen transfer, mixing in air-lift reactor) |
| Medium | Molasses or glucose syrup plus additional nutrients and salts 150 kg m$^{-3}$ 140–220 kg m$^{-3}$ | | up to 280 kg m$^{-3}$ |
| Medium pre-treatments | Low manganese concentration requiring medium pre-treatment with HCF or copper ions | | No metal ion pre-treatment required |
| Other features | NH$_4^+$ stimulates citric acid production | Mycelial morphology as pellets | Nitrogen limitation triggers acid accumulation. Thiamine required for acid accumulation |

*vvm = volume of air per unit volume of medium per minute.

production involves oxygen consumption, the rate of acid production increases with increased level of dissolved oxygen. In addition, a short interruption of the oxygen supply can lead to irreversible cessation of citrate production. It is essential to maintain the pH below 2.0 for citrate biosynthesis. At higher pH values, *A. niger* accumulates gluconic acid rather than citrate.

Citrate production by *Candida guilliermondii* differs in a number of ways from the *A. niger* submerged process. Manganese deficiency is not a prerequisite and a metal-removal medium pre-treatment step is unnecessary. Citrate is produced at a high pH (3.5–5.0). In addition, nitrogen limitation triggers acid accumulation. The main advantages of the *Candida* citrate process over the *A. niger* process relate to its higher overall fermentation productivity. It is possible to use higher sugar concentrations because of the more osmotolerant nature of the organisms. In addition, the fermentation is faster.

*Biochemistry of citrate production by A. niger*
The metabolic process leading to citrate accumulation involves (a) breakdown of hexoses to pyruvate and acetyl CoA, (b) the anaplerotic formation of oxaloacetate from pyruvate and $CO_2$, and (c) the accumulation of citrate within the tricarboyxlic acid cycle. In this scheme, no $CO_2$ should be expended during active citrate production as $CO_2$ released during oxidative decarboxylation of pyruvate to acetyl-CoA should be utilized in the conversation of pyruvate to oxaloacetate. The key enzyme, pyruvate carboxylase, catalysing this reaction, is constitutively produced in *Aspergillus* species. High sugar concentrations further increase the activity of this enzyme as well as glycolytic enzymes and may repress some of the tricarboyxlic acid cycle enzymes. In order for citrate to accumulate, it is necessary that at least one of the enzymes of the tricarboxylic acid cycle is inhibited. Recent evidence suggests the major regulatory step is α-ketoglutarate dehydrogenase, a rate-limiting step in the cycle and the only irreversible reaction. This enzyme is inhibited by the increased physiological concentrations of oxaloacetate and NADH which occur during citrate production. Phosphofructokinase is inhibited by citrate, but this inhibition can be counteracted by elevated intracellular concentrations of $NH_4^+$.

Manganese deficiency reduces the activity of some enzymes of the pentose phosphate pathway (which would divert hexoses away from glycolysis and citrate production) and also inhibits the tricarboxylic acid cycle. Manganese deficiency impairs anabolic metabolism in general including protein and nucleic acid turnover. During manganese deficiency, an acid protease is formed and intracellular pools of nucleic acids and proteins decrease with concomitant production of peptides, amino acids and elevated levels of $NH_4^+$. It has been concluded that the main effect of manganese deficiency relates to its impact on protein turnover, causing the increased concentrations of $NH_4^+$ necessary to counteract phosphofructokinase inhibition by citrate.

Oxygen is required for the metabolic re-oxidation of NADH during citric acid production. During citrate accumulation, an alternative respiratory system, sensitive to salicylhydroxamic acid (SHAM) supports the standard respiratory

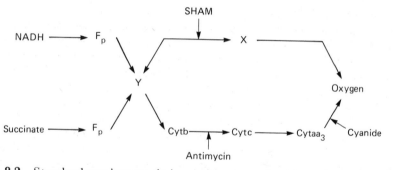

**Fig. 8.2** Standard respiratory chain sensitive to antimycin and cyanide and alternative SHAM-sensitive system in *A. niger*.

system in NADH re-oxidation, but without concomitant ATP production (Fig. 8.2). The fact that citric acid accumulation is strongly inhibited by SHAM illustrates the importance of this system. SHAM-sensitive respiration depends on a high oxygen tension and the system is inactivated by a short interruption in aeration.

Gluconic acid production by *A. niger* is catalysed by an extracellular, partially mycelium-bound, glucose oxidase. This enzyme is inactivated below pH 2.0. At higher pH values, gluconic acid is produced from glucose by *A. niger* and glucose induces this enzyme above pH 4.0—hence the requirement to operate citrate production fermentations below pH 2.0.

In summary, citrate production by *A. niger* is characterized by the following:

- carbohydrate-induced high activity of glycolytic enzymes and pyruvate carboxylase leading to citrate synthesis
- inhibition of a tricarboxylic acid cycle enzyme which would cause citrate breakdown
- manganese deficiency-mediated production of an intracellular $NH_4^+$ pool to counteract citrate inhibition of phosphofructokinase
- high oxygen tension and continued oxygen supply to maintain SHAM-sensitive respiration active for re-oxidation of NADH
- low pH to inactivate glucose oxidase.

GLUCONIC ACID PRODUCTION

D-Gluconic acid, gluconate salts and D-glucono-δ-lactone are non-toxic chemicals which may be produced from glucose by electrochemical or catalytic oxidation or by fermentation using *A. niger* or *Gluconobacter suboxydans*. Fermentation is currently the preferred method. D-Glucono-δ-lactone is used as a latent acid in baking powders. Sodium gluconate is used in the presence of sodium hydroxide as a sequestering agent of calcium in glass-bottle washing, and of iron in alkaline de-rusting of ferrous metals. Sodium gluconate is also used in cement mixes. Calcium gluconate and iron(II) gluconate are used pharmaceutically in the treatment of calcium and iron deficiency.

*Gluconate fermentation process*

The medium for sodium gluconate production by *A. niger* consists of approximately $250 \, g \, l^{-1}$ starting glucose concentration, ammonium salts, urea or corn steep powder as nitrogen source and other nutrients. Initial glucose level may be supplemented by glucose feeding up to a total level of $600 \, g \, l^{-1}$. Too much nitrogen leads to excessive growth and decreased acid yields. The production pH is maintained at 6.0–7.0 using NaOH until optimum growth and glucose oxidase level has been achieved, at which stage the pH may be allowed to drop to 3.5. The fermentation has a high oxygen requirement and temperature is controlled at 30–33°C. Yield of product exceeds 90% of theoretical yield. Typical industrial fermentations can produce sodium gluconate at an average rate of $10–13 \, g \, l^{-1} \, h^{-1}$ for a fermentation period of 20–60 h. However, because product crystallization problems may be encountered with final sodium gluconate concentrations of $600 \, g \, l^{-1}$, obtained in more prolonged fermentations, a shorter fermentation duration may be chosen.

For calcium gluconate production, calcium carbonate is used to neutralize gluconic acid production and maintain the fermentation pH above 3.5. In order to avoid precipitation of calcium gluconate, the amount of calcium carbonate added is only about two-thirds the stoichiometric requirement. The additional amount required for complete neutralization is added to the filtered broth after fermentation and calcium gluconate crystals are recovered.

Free gluconic acid is predominantly prepared from sodium gluconate by ion-exchange methods. Gluconic acid exists in aqueous solution in equilibrium with glucono-δ-lactone and glucono-γ-lactone, with the ratios being dependent on concentration and temperature. Crystals of gluconic acid, glucono-δ-lactone and glucono-γ-lactone separate from over-saturated solutions at 0–30°C, 30–70°C and above 70°C, respectively, and this enables commercial glucono-δ-lactone to be recovered. Commercial gluconic acid is sold as a 50% aqueous solution.

*Biochemistry of gluconate production*

α-D-Glucose is converted to β-D-glucose spontaneously and in *A. niger* the reaction is accelerated by the enzyme mutarotase. β-D-Glucose is converted to D-glucono-δ-lactone, mediated by the glucose oxidase of *A. niger*. Glucose oxidase, a flavoprotein, is reduced by removing two hydrogens from glucose. The flavoprotein is then re-oxidized by molecular oxygen, giving $H_2O_2$ which is decomposed by catalase. The two enzymes are contained within peroxisomes, thereby preventing $H_2O_2$ cell toxicity during gluconate production. As already mentioned, glucose oxidase is induced by glucose at pH values above 4.0 and is denatured at pH 2.0. The conversion of β-D-glucose to D-glucono-δ-lactone by *Gluconobacter suboxydans* is mediated by the enzyme NADP glucose dehydrogenase.

The equilibrium existing between D-glucono-δ-lactone and D-gluconic acid has already been noted and conversion to gluconic acid occurs at neutral pH. This spontaneous conversion is less effective at low pH values and is facilitated in some processes involving *A. niger* by the mediation of a D-glucono-δ-lactonase. Since glucono-δ-lactone accumulation has a negative effect on the rate of glucose oxidation, these processes for its efficient removal are important. While metabolic

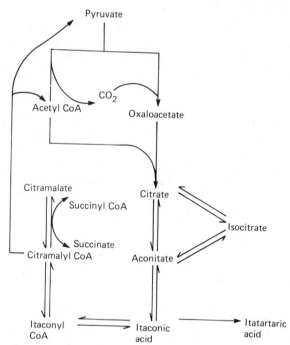

**Fig. 8.3**   Metabolic pathway involved in itaconic acid production.

pathways exist for catabolism of the gluconic acid formed, breakdown appears to be retarded in industrial gluconate strains under process conditions such as excess glucose concentration and the prevailing pH.

## ITACONIC ACID

Itaconic acid is a valuable intermediate for polymer chemistry because of its unique combination of two carboxyl groups and a methylene group. Itaconic acid itself polymerizes to give only low-molecular-weight polymers and consequently it functions best in co-polymers. It is also used in synthesis of pyrrolidones and as an emulsion paint additive.

Itaconic acid is produced by aerobic submerged culture of *Aspergillus terreus* in media containing molasses and ammonium salts or corn steep as carbon and nitrogen sources, respectively. Optimum growth of the organism is achieved at pH 5–7 whereas a lower pH of 3–4 is optimum for itaconic acid production. Initial sugars range from $100–180\,g\,l^{-1}$ and the yield of product is 55–65% based on carbohydrate weight. Fermentation time is about 72 h. A short aeration break can irreversibly stop itaconic acid production. Itaconic acid is produced from citrate via aconitate (Fig. 8.3). The metabolism of hexose to citrate is similar to that involved in *A. niger* citrate production.

## LACTIC ACID

About half of the world demand for lactic acid is produced by fermentation. Its main industrial uses are as a food acidulant (50% of market), for manufacture of stearoyl-2-lactylate (20%) and in pharmaceutical and other applications.

L-(+)-Lactic acid is produced in anaerobic fermentation using *Lactobacillus delbruckii* and related homofermentative strains. Media contain about 15% sucrose or dextrose and complex nitrogen. pH is controlled in the region 5.0–6.5 by $CaCO_3$ or $Ca(OH)_2$ neutralization, temperature is maintained between 45–60°C and fermentation time is 3–4 days. Yields of 90–95% lactic acid, based on initial sugar content, are obtained. The fermentation process involves conversion of hexose to pyruvate via the Embden–Meyerhof–Parnas pathway and its conversion to L-(+)-lactate by the enzyme L-lactate dehydrogenase.

## GLYCEROL

Glycerol was produced by fermentation as a material for explosives manufacture during World Wars I and II.

Glycerol is formed by yeast in tiny amounts along with ethanol in the alcoholic fermentation. Normally during alcohol production, $NADH + H^+$, formed during glycolysis by conversion of glyceraldehyde-3-phosphate to 1,3-diphosphoglyceric acid, is re-oxidized during conversion of acetaldehyde to ethanol. Sodium bisulphite addition, however, results in the formation of an acetaldehyde–sulphite complex and the $NADH + H^+$ is available for reduction of the glycolytic intermediate dihydroxyacetone phosphate to glycerol phosphate, which is dephosphorylated to glycerol. Because other by-products are also formed, yields never exceed 30%, based on carbohydrate weight, in a 2–3 day fermentation. The metabolic process is summarized in Fig. 8.4.

## ACETONE–BUTANOL

Acetone is used as a solvent in manufacture of lacquers, resins, rubbers, fats and oils and butanol has applications in the production of lacquers, rayon, detergents, brake fluids, amines and as a general solvent. An acetone–butanol fermentation has been successfully operated commercially for many years, and the last production facility, run by National Chemical Products SA, was closed only a few years ago.

This process involves anaerobic fermentation of molasses, starch or cruder materials, at a carbohydrate concentration of 5–6.5%, by strains of *Clostridium acetobutylicum*, to yield about a 2% concentration of solvent mixture. The fermentation may be characterized as having three phases. Phase 1 involves 12–14 h of rapid growth, production of acetic and butyric acid, a pH decline from 6.0 to about 4.0 and evolution of $CO_2$ and $H_2$. Phase 2 involves conversion of acids to solvent with increased $CO_2$ production and a reduction of titratable acidity. In

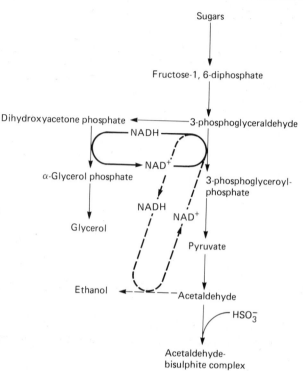

**Fig. 8.4** Production of glycerol by a modification of the yeast ethanol fermentation.

phase 3, gas and solvent production declines and cells begin to autolyse. Further development of this fermentation is being attempted to regulate metabolism to produce predominantly one solvent product, to reduce the toxic effect of products on cell physiology and to increase overall production rate and efficiency in an effort to make this process commercially viable.

ACETIC ACID

The annual world market for acetate is currently 2.5 million t. Synthetic methods have provided the major proportion of the world's acetic acid supply since 1950. Since ethylene is the main raw material used, and this increased in price from 6 cents/kg to 30 cents/kg from 1970 to 1980, alternative production routes, including biological methods, may assume importance. Brazil produces acetic acid by chemical conversion of ethanol, derived solely from biomass. Fermentation methods for production of chemical feedstock acetate also have potential.

The aerobic process for production of vinegar has already been described (see Chapter 7). In anaerobic acidogenic fermentations, a variety of volatile fatty acids are produced. Acetic acid constitutes the largest component of the acid mixture, with significant amounts of propionic and butyric acid also being produced. *Clostridium butyricum* produces a mixture of acetic acid and butyric acid from

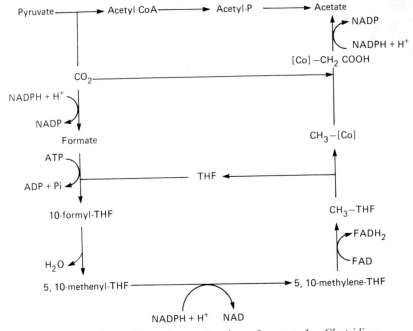

**Fig. 8.5** Metabolic pathway for production of acetate by *Clostridium thermoaceticum*.

glucose. *Clostridium thermocellum* and *Clostridium thermoaceticum* ferment glucose and fructose almost stoichiometrically to acetate. In acidogenesis, it is imperative to suppress methanogens, often associated with acid formers in mixture culture, which would convert the acid to methane and $CO_2$.

While a higher acetate concentration may be obtained with aerobic processes, the theoretical yield of acetate from anaerobic acidogenic processes is 1.0 compared to 0.66 for aerobic processes. The $2CO_2$ molecules produced during conversion of pyruvate to acetyl CoA are assimilated to account for the third acetate molecule formed from glucose (Fig. 8.5). Furthermore, energy requirements for the anaerobic process are lower and low value substrates, such as distillery wastes and sulphite waste liquor, may be used. Consequently, anaerobic acetogenic fermentations may become the method of choice for production of chemical feedstock acetate.

## 2,3-BUTANEDIOL

Although less expensive chemical routes currently exist for manufacture of 2,3-butanediol from petroleum, alternative fermentation routes are considered to have significant potential. *Klebsiella oxytoca* yields 2,3-butanediol as major product from xylose as well as glucose and therefore offers an opportunity for microbial conversion of low value substrates, derived from wood hydrolysate, to product.

*Bacillus polymyxa* also utilizes both pentoses and hexoses to produce 2,3-butanediol and ethanol in equal amounts.

Although 2,3-butanediol is a product of anaerobic metabolism, a limited controlled oxygen supply is required to increase cell density. Fermentation products excreted by *K. oxytoca*, in addition to 2,3-butanediol, include acetoin, ethanol and acetate and the oxygen supply rate can control the proportions of the various metabolites produced. Equimolar amounts of ethanol and 2,3-butanediol are produced in the absence of oxygen. Limited oxygen supply inhibits ethanol production and maximizes 2,3-butanediol production. A further increase in oxygen supply shifts metabolism completely from fermentation to respiration, with total conversion of substrate to cell mass and $CO_2$. Consequently, the most important variable affecting rate and yield of 2,3-butanediol production is oxygen availability.

The maximum theoretical yield of 2,3-butanediol produced from glucose is 0.5 and actual yields are 80–90% of theoretical. Disadvantages in potential use of *K. oxytoca* include its low osmotic tolerance and problems of substrate and product inhibition. Under a variety of batch and continuous experimental fermentation conditions, final concentrations of 2,3-butanediol ranged from 30–99 g l$^{-1}$ with productivities ranging from 0.9–3.0 g l$^{-1}$ h$^{-1}$.

2,3-Butanediol production from hexoses and pentoses involves formation of pyruvic acid as key metabolic intermediate. Two pyruvate molecules are condensed to form acetolactate and a decarboxylase converts this to acetoin. Acetoin may be oxidized to 2,3-butanedione (diacetyl) or enzymatically reduced to 2,3-butanediol by acetoin reductase.

## Gibberellic acid

Gibberellins are one class among the five known classes of phytohormones. Among a large family of gibberellins, gibberellic acid (Fig. 8.6), used to accelerate malt production, is the most important compound commercially. It is produced by cultivation of *Fusarium moniliforme*, the imperfect stage of the fungus *Gibberella fujikuroi*. Surface culture processes were originally used, producing yields of 40–60 mg/l gibberellic acid in a prolonged 2–3 week fermentation. Commercial fermentations are now carried out by submerged culture, where yields of 1–2 g/l can be achieved in about 6 days. Two important fermentation media conditions for gibberellin production are a low nitrogen content and a mixture of carbon sources. The fermentation has been characterized as having six physiologically distinct stages (Fig. 8.7): (1) Lag phase; (2) growth phase without nitrogen

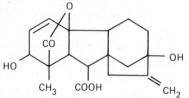

**Fig. 8.6** Gibberellic acid.

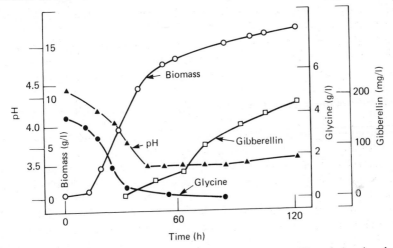

**Fig. 8.7** Glycine-limited batch fermentation for gibberellin production by *Gibberella fujikuroi* (reproduced with permission from Bu'Lock *et al.*, 1974).

limitation with low gibberellin production; (3) glycine limitation, low cell-growth rate, slow gibberellin production; (4) glycine depleted, residual glucose catabolized, high rate of gibberellin formation; (5) glucose depleted, reduced gibberellin accumulation; (6) cell lysis, pH increase.

## Biopolymers

In this section production of two types of biopolymers, microbial polysaccharides and poly-$\beta$-hydroxybutyrate, will be considered.

### MICROBIAL POLYSACCHARIDES

Traditionally, industrial polysaccharides have been recovered from plants and marine algae. Microbial polysaccharides provide a viable alternative as replacement products for some of these existing plant or algal materials but perhaps more importantly may be used in a wide range of new industrial applications. Genetic mutation or engineering of some of the producing organisms and/or environmental manipulation of fermentation conditions provide the potential for manipulation of product properties and specification, thereby extending the diversity of available biopolymers.

Xanthan gum is the only fermentation-based microbial polysaccharide which has a significant share of the world polysaccharide market. Its basic repeating unit consists of a pentasaccharide containing glucose, mannose, glucuronic acid, acetate and pyruvate. This gum has a high viscosity in low concentration which is stable over a wide pH range, is independent of temperature and of the presence of cations. In combination with plant polysaccharides in aqueous solutions, it forms stable gels. It has applications as a stabilizer of gel suspensions and for viscosity control. Because of its pseudoplastic properties, combined with its stability to temperature and cations, it has advantages as a lubricant. It is used in combination with surfactants and hydrocarbons in enhanced oil recovery.

Alginates are obtained from seaweeds, but since this source is subject to much variability, there is considerable industrial interest in replacing it with alginate-like polysaccharides, produced by fermentation, such as the polysaccharide of *Azotobacter vinelandii*. Other microbial polysaccharides of commercial interest include pullulan, produced by *Aureobasidium pullulans*, seleroglucan, produced by *Selerotium* species and gellan, produced by *Pseudomonas elodea*.

Processes for microbial exopolysaccharide production are characterized by the high viscosity of the fermentation medium, the low product concentrations achieved and polymer conformational changes which occur over the course of the fermentation. Control of medium constituents and other fermentation parameters is critical to achieve the desired rates of synthesis. Limited studies indicate that modification of exopolysaccharide, by manipulation of the growth medium, appears to affect the degree of polymerization rather than the actual structure and composition of the repeating unit.

In the case of xanthan gum production, the content of pyruvate as substituent can vary from 0–75% by appropriate strain and medium selection, and this can effect the polymer's rheological properties, when polymer concentration changes or when medium ionic strength or temperature is altered. Some growth media for xanthan production result in the formation of non-polysaccharide producing mutants, but this can be avoided by proper choice of medium and by using nutrient-limited conditions for growth. Generally with charged microbial polysaccharide production, pH control is desirable. Optimum pH for xanthan production is 7.0, and growth and product formation ceases at pH 5.5. Over the course of the fermentation, the properties of the broth change from low-viscosity Newtonian conditions to highly-viscous non-Newtonian conditions, as exo-polysaccharide concentration increases. These changes have a profound effect on the mixing, mass and heat-transfer properties of the fermentation, and fermenter design and operation must be optimized to suit these rheological conditions. Reported final concentrations of xanthan in fermenter broths are around $50 \, g \, l^{-1}$ and yields based on glucose consumed are 50–60%.

Most bacterial exopolysaccharide synthesis probably occurs intracellularly, involving utilization of the activated sugar nucleotide pathway. Sugars are subsequently transferred from the nucleotide to a lipid carrier, which has been activated by initial transfer of a sugar phosphate, resulting in the formation of the sugar repeating unit. The exact method of chain elongation and polysaccharide extension is not understood.

## POLY-$\beta$-HYDROXYBUTYRATE (PHB)

The potential major application of poly-$\beta$-hydroxybutyrate (PHB) is as a substitute for plastics in cases where its properties of biodegradability would be an advantage. The ability of bacteria to re-oxidize NADH becomes limiting at low oxygen concentrations. Re-oxidation is then achieved by producing products such as PHB, ethanol or butanol. PHB is accumulated in this way as a reserve material by a wide variety of bacteria and *Alcaligenes eutrophus* can accumulate 70–80% of its biomass as PHB.

The repeating unit or monomer in PHB, $\beta$-hydroxybutyrate, is synthesized from two acetyl CoA units, and the desired molecular weight of the polymer should be 200 000–300 000. Synthesis of PHB is promoted by phosphorus or nitrogen limitation in addition to being influenced by oxygen concentration. The fermentation process for commercial PHB production would most likely have two stages, a growth stage, without substrate or oxygen limitation, to accumulate high biomass concentrations, followed by a product formation stage under optimized conditions of nutrient limitation. While glucose is currently used as substrate by *A. eutrophus*, the possibility of using cheaper substrates which would significantly reduce production costs would be commercially desirable.

## Bioinsecticides

Major reservations concerning the use of chemical insecticides relate to (a) their lack of specificity and the possibility that non-target insects may also be killed, (b) the potential development of target organisms resistant to the insecticide and (c) in some cases their undesirable persistance in the environment. Microbial insect control involves the use of spores or vegetative cells of insect pathogenic micro-organisms. Over 400 species of fungi and greater than 90 bacterial species are known to attack insects and mites.

*Bacillus thuringiensis* varieties, used to control caterpillars, mosquitoes and blackflies, are now widely established bulk fermentation products and represent about 90% of the total bioinsecticide market. Sporulated cells of *B. thuringiensis* contain a protein crystal of high molecular weight called the delta-endotoxin, toxic to lepidopteran larvae. The crystal is digested by proteases, in the alkaline mid-gut of caterpillars, into toxic sub-units which appear to cause disintegration of gut surface epithelial cells, resulting in paralysis and death. *B. thuringiensis* also produces a second toxin, beta-exotoxin or thuringiensin toxic to house flies. Many sub-species exist and isolates of the sub-species *kurstaki* do not contain the beta-exotoxin. *B. thuringiensis* var. *israelensis* is relatively specific against aquatic diptera larvae such as mosquitoes. *Bacillus sphaericus*, which also produces a potentially useful insecticide for mosquito control and *Bacillus popilliae*, an insect pathogen of the Japanese beetle, both have significant commercial potential, provided difficulties of *in vitro* production by fermentation can be overcome.

*Bacillus thuringiensis* is produced by aerobic submerged cultivation in cheap complex media, containing ingredients such as soybean meal, corn starch, corn steep, molasses, yeast extract and hydrolysed casein. Fermentation conditions are optimized to achieve high bacterial growth rates and cell yields and efficient sporulation with concomitant crystal toxin production.

Fungi generally infect insects through the cuticle, with invasion dependent on production of chitinases, proteases and lipases. The major entomopathogenic fungi are Deuteromycotina including *Beauveria bassiana* and *Metarrhizium anisopliae* (general pathogens), *Verticillium lecanii* (a pathogen of white flies and aphids), and *Hirsutella thompsonii* (a pathogen of eriophyid mites). These species are primarily of interest because they grow *in vitro*. In some fungi, the conidia are the agents responsible for infection and often solid or semi-solid cultivation techniques are required. However, in the case of *B. bassiana* and *H. thompsonii*, conidial formation and infective spore production has been achieved in submerged culture. Several species, including *V. lecanii*, assume a yeast-like morphology in submerged culture, forming budding elements or blastospores. Blastospore production has not been widely used commercially because blastospores tend to be unstable and have low infectivity. If these problems could be resolved submerged fermentation processes would be the most effective method of fungal entomopathogen production. In many cases the nutritional requirements of insect pathogenic fungi are complex and poorly understood, and there is much scope for research leading to optimization of culture conditions for production of fungal insecticides.

Insect viruses can be produced by fermentation techniques by viral infection of susceptible insect cells grown in cell culture. Many insect cell lines, especially from lepidoptera and diptera species, have now been established and a variety of different techniques have been used to optimize cell growth and viral replication in insect cell cultures. Technical problems associated with large-scale cell-culture techniques and the high cost of cell-culture media are currently major barriers to the development of commercially-viable fermentation processes for production of viral insecticides.

# Chapter 9

# *Food additives*

In this chapter fermentation processes for production of some important food additives, amino acids, nucleotides, vitamins, fats and oils will be discussed. The distinction between food additives and industrial chemicals is not all that clear-cut and processes for production of some chemicals, such as citric acid and microbial polysaccharides, which have non-food uses in addition to food additive applications, have already been described in Chapter 8. Single-cell protein products and vinegar (Chapter 7), and many enzymes (Chapter 11) may also be considered as food additives. Likewise some of the food additives discussed in this chapter have other applications. For example, glutamic acid is used as the starting material for synthesis of specialty chemicals such as N-acylglutamate, a bio-degradable surfactant, and oxopyrrolidinecarboxylic acid, a natural moisturizing factor. Similarly $\gamma$-linolenic acid is used as a prostaglandin precursor.

## Amino acids

Amino acids are produced using a range of technologies including direct fermentation, biotransformation of precursors using cells or enzymes, extraction of protein hydrolysates and chemical synthesis. They have a variety of uses as nutrients and flavours in the food and feed industries. Table 9.1 indicates the annual demand, production methods and applications of amino acids in the food industry. Important amino acids with non-food applications include L-arginine, L-glutamine, L-histidine, L-leucine, L-phenylalanine, L-tyrosine and L-valine.

While fermentation or biotransformation processes have been developed for production of all amino acids except glycine, L-cysteine and L-cystine, not all of these processes are commercially viable. L-Asparagine, L-leucine, L-tyrosine,

**Table 9.1** Annual production of amino acids with food applications

| Amino acids | Annual production (metric tons) | Production methods | | | | Major food/feed application |
|---|---|---|---|---|---|---|
| | | Direct fermentation | Biotransformation enzymes/cells | Extraction of protein hydrolysates | Chemical synthesis | |
| L-Alanine | 50 | | + | | | Flavour enhancer |
| D,L-Alanine | 200 | | | | + | Flavour enhancer |
| L-Aspartate | 1 000 | | + | | | Flavour enhancer |
| L-Cysteine | 200 | | | + | | Baking antioxidant |
| L-Glutamate | 400 000 | + | | | + | Flavour enhancer |
| Glycine | 6 000 | | | | + | Sweetener component |
| L-Lysine | 40 000 | + | | | | Feed additive |
| D,L-Methionine | 70 000 | | | | + | Feed additive |
| L-Threonine | 100 | + | | | + | Feed additive |

L-cysteine and L-cystine are produced by purification of protein hydrolysates. Chemical synthesis is more economical for production of optically-inactive racemic mixtures of D- and L-isomers, and D,L-alanine, D,L-methionine, D,L-tryptophan and glycine are produced in this way. Processes involving amino acylase enzymes may be used to resolve these racemic mixtures (see Chapter 11).

Microbial strains from the genera *Corynebacterium* and *Brevibacterium* have assumed major importance in the production of amino acids by fermentation. Natural isolates of these strains can excrete large quantities of glutamic acid. Because of cell metabolic regulatory mechanisms, particularly end-product repression and inhibition, substantial levels of amino acids are rarely excreted by wild-type isolates. Production of commercial quantities of the amino acids has been dependent on the successful development of deregulated mutants. The two most important methods involve use of auxotrophic and regulatory mutants or a combination of the two. Auxotrophic mutants, which lack the enzyme needed to form the regulatory effector metabolite (often the end-product), may accumulate and excrete the metabolic intermediate which is the substrate for the eliminated enzyme. A lysine auxotroph, for example, lacks an enzyme in the pathway necessary for lysine synthesis and requires lysine, or a metabolic precursor which can be converted to lysine, for growth. End-product inhibition by the amino acid product of an unbranched biosynthetic pathway may be avoided by the development of regulatory mutants, having an altered feedback-insensitive key enzyme, thus allowing accumulation of the particular amino acid. Analogues of the end-product, which are also capable of inhibition of the sensitive key enzyme, may be used in screening methods for selection of analogue-resistant or regulatory mutants. Revertants may be selected from auxotrophic mutants (apparently lacking the key regulatory enzyme) which produce a modified deregulated enzyme. Table 9.2 illustrates the genetic features of mutants of *Brevibacterium* spp. and *Corynebacterium* spp. and some published yields of amino acids over-produced from glucose.

GLUTAMIC ACID PRODUCTION

The flavour-enhancing properties of sodium glutamate were discovered in Japan at the start of the twentieth century and a fermentation process for its production by *Corynebacterium glutamicum* currently supplies an annual world market of about 400 000 t.

Molasses or starch hydrolysates are generally used for commercial production of glutamic acid by *C. glutamicum* and related strains. An ample supply of a suitable nitrogen source such as ammonium salts is essential, since $NH_3$ is incorporated into the amino acid molecule. Glutamic acid producing bacteria can also utilize urea as nitrogen source. The concentration of ammonium ion must be maintained at a low level in the medium as higher concentrations are detrimental to cell growth and product formation. The pH of the medium tends to drop due to cell glutamate excretion and ammonium ion assimilation, and gaseous ammonia is used as a means of simultaneously controlling medium nitrogen level and a

**Table 9.2** Genetic characteristics of some amino acid-producing strains of *Brevibacterium flavum* and *Corynebacterium glutamicum*

| Microbial strain | Amino acid | Genetic characteristics | Yield (g l$^{-1}$) |
|---|---|---|---|
| *Brevibacterium flavum* | | | |
| | L-Arginine | Gua$^-$TA$^r$ | 35 |
| | L-Histidine | TA$^r$SM$^e$Eth$^r$ABT$^r$ | 10 |
| | L-Isoleucine | AHV$^r$OMT$^r$ | 15 |
| | L-Lysine | AEC$^r$ | 57 |
| | L-Proline | Ile$^-$ SG$^r$DHP$^r$ | 29 |
| | L-Threonine | Met$^-$ AHV$^r$ | 18 |
| *Corynebacterium glutamicum* | | | |
| | L-Glutamate | Wild type | >100 |
| | L-Glutamine | Wild type | 40 |
| | L-Lysine | Hom$^-$ Leu$^-$ AEC$^r$ | 39 |
| | L-Phenylalanine | Tyr$^-$ PFP$^r$PAP$^r$ | 9 |
| | L-Tryptophan | Phe$^-$Tyr$^-$ 5MT$^r$TrpHx$^r$6FT$^r$ 4MT$^r$PFP$^r$PAP$^r$TyrHx$^r$PheHx$^r$ | 12 |
| | L-Tyrosine | Phe$^-$ PFP$^r$PAP$^r$PAT$^r$TyrHx$^r$ | 18 |

Resistance abbreviations: r, resistant; ABT, 2-aminobenzthiazole; AEC, S-($\beta$-aminoethyl)-L-cysteine; AHV, $\alpha$-amino-$\beta$-hydroxyvaleric acid; DHP, 3,4-dehydroproline; Eth, ethionine; 6FT, 6-fluorotryptophan; 4MT, 4-methyltryptophan; 5MT, 5-methyltryptophan; OMT, O-methylthreonine; PAP, *p*-aminophenylalanine; PheHx, phenylalanine hydroxamate; SG, sulfaguanidine; TA, 2-thiazolalanine; TyrHx, tyrosine hydroxamate; TrpHx; tryptophan hydroxamate.

Auxotroph abbreviations: $-$, auxotroph; Ile, isoleucine; Met, methionine; Leu, leucine; Hom, homoserine; Phe, phenylalanine; Tyr, tyrosine; Gua, guanine.

fermentation pH optimum of 7.0–8.0. Glutamate biosynthesis is an aerobic process requiring oxygen throughout the fermentation.

Glutamate-producing bacteria require biotin for growth but accumulation of the amino acid is maximum at a critical biotin concentration of $0.5\,\mu g/g$ cells (dry), which is sub-optimal for maximum growth. Excess biotin, while supporting abundant growth, impairs glutamate accumulation. Addition of $C_{16}$–$C_{18}$ saturated fatty acids during growth also permits accumulation of glutamate even in the presence of high biotin concentrations. This is because accumulation of the amino acid is primarily controlled by its rate of excretion rather than by its rate of biosynthesis. Biotin is a co-factor of acetyl CoA carboxylase, the first enzyme in the pathway for biosynthesis of oleic acid (unsaturated, $C_{18:1}$) and its subsequent incorporation into phospholipids. $C_{16}$–$C_{18}$ saturated fatty acids repress acetyl CoA carboxylase. Phospholipids appear to regulate the permeability of the cell to glutamate and sub-optimal biotin concentrations or $C_{16}$–$C_{18}$ saturated fatty acids act to decrease the concentration of phospholipids in the cell, thereby increasing cell permeability to glutamate.

Even in the presence of excess biotin, glutamic acid producing bacteria, grown in the presence of penicillin, can accumulate large amounts of glutamate. Penicillin inhibits bacterial cell-wall synthesis and the enhanced accumulation of glutamate is thought to result from the formation of swollen cells with weakened cell walls, resulting in damage to the permeability barrier of the cell membrane.

Under optimized conditions for glutamate production from hexose, the Embden–Meyerhof–Parnas pathway predominates, channelling carbon precursors into the tricarboxylic acid cycle. The $NADPH + H^+$ formed in the oxidative decarboxylation of isocitrate to $\alpha$-ketoglutarate provides the reduced co-factor which together with $NH_3$ is required for conversion of $\alpha$-ketoglutarate to glutamate by glutamate dehydrogenase. Commercial glutamic acid-producing strains lack the tricarboxylic acid cycle enzyme $\alpha$-ketoglutarate dehydrogenase and consequently, in the absence of $NH_4^+$ ions but with sufficient oxygen, $\alpha$-ketoglutaric acid accumulates. Krebs cycle intermediates, required for replenishment of oxaloacetate needed for the citrate synthase condensation reaction and other cell reactions, are supplied by efficient anaplerotic reactions. Phosphoenolpyruvate carboxylase plays an important role in the carboxylation of phosphoenolpyruvate to form oxaloacetate. Alternatively, the Krebs cycle intermediates may be replenished by the operation of the glyoxylate cycle (see Chapter 2). Stoichiometrically 1 mole of glutamate is produced from 1.4 mole glucose when the glyoxylate cycle is involved whereas the pathway involving carbon dioxide fixation by phosphoenolpyruvate carboxylase is more efficient, producing 2 moles of glutamate per mole of glucose. In order to increase the efficiency of conversion, some mutants have been introduced which have decreased levels of the glyoxylate cycle enzyme, isocitrate lyase. The metabolic pathway for production of glutamate from glucose is summarized in Fig. 9.1.

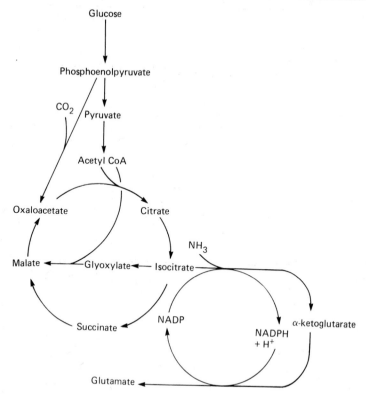

**Fig. 9.1** Metabolic pathway for production of glutamic acid from glucose.

## LYSINE PRODUCTION

Lysine, an amino acid which is essential for animal and human nutrition, is lacking in cereals and this has created an annual market for lysine in excess of 40 000 t. The amino acid is produced predominantly by direct fermentation using an auxotrophic mutant of *Corynebacterium glutamicum*. Recently, a second fermentation process, involving a regulatory mutant of *Brevibacterium flavum* and a biotransformation process, involving conversion of chemically-synthesized α-aminocaprolactam to L-lysine, have been commercialized.

The pathway for biosynthesis of lysine by *C. glutamicum* and *B. flavum* is illustrated in Fig. 9.2. The first enzyme, aspartokinase, is regulated by concerted feedback inhibition by L-threonine and L-lysine. L-Threonine causes feedback inhibition of homoserine dehydrogenase, while L-methionine represses synthesis of this enzyme. Hence, a homoserine auxotroph or a threonine–methionine double auxotroph of *C. glutamicum* diminishes the intracellular pool of threonine and reduces its marked feedback inhibitory effect on aspartokinase and promotes good lysine production. S-(2-aminoethyl)-L-cysteine, a lysine analogue (SAEC)

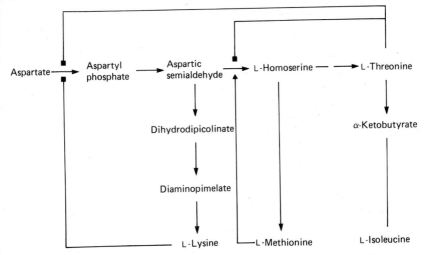

**Fig. 9.2** Metabolic pathway for lysine biosynthesis in *Corynebacterium glutamicum* and *Brevibacterium flavum*. ■, inhibition; ▲, repression.

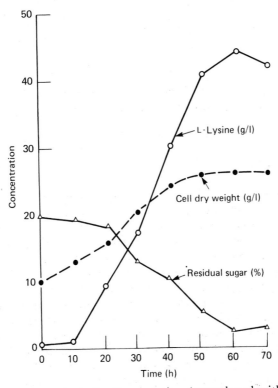

**Fig. 9.3** Time course of a lysine fermentation (reproduced with permission from Nakayama, 1985).

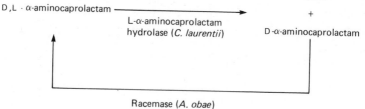

**Fig. 9.4**   Conversion of D,L-α-aminocaprolactam to L-lysine by biotransformation.

which behaves as a false feedback inhibitor of aspartokinase, inhibits growth of *B. flavum*. Growth inhibition is markedly enhanced by L-threonine and is reversed by L-lysine. Some mutants, capable of growth in the presence of both SAEC and L-threonine and considered to contain aspartokinase which has been desensitized to concerted feedback inhibition, are potent L-lysine producers. Combined regulatory and auxotrophic mutants, resistant to SAEC and requiring homoserine or threonine and methionine for growth, over-produce lysine.

Biosynthesis of aspartate for lysine production is from the oxaloacetate component of the Krebs cycle. The predominant anaplerotic reaction is considered to be the phosphoenolpyruvate carboxylation system rather than the glyoxylate cycle. Incorporation of excess biotin into the medium inhibits undesirable glutamate over-production.

An example of a fermentation profile for production of lysine by a homoserine auxotroph is illustrated in Fig. 9.3.

D,L-α-Aminocaprolactam, chemically synthesized from cyclohexane, is the raw material for production of L-lysine by biotransformation. Acetone-dried cells of *Cryptococcus laurentii*, which contain the enzyme L-α-aminocaprolactam hydrolase, convert L-aminocaprolactam to L-lysine, while acetone-dried cells of *Achromobacter obae* contain a racemase to convert D-α-aminocaprolactam to the L-form (Fig. 9.4).

## Nucleosides

The flavour-enhancing effects of katsubushi, used in Japan, is due to the histidine salt of 5'-inosine monophosphate (IMP). 5'-Guanosine monophosphate (GMP) also exhibits strong flavour-enhancing properties. There is also synergism between the flavour-enhancing properties of IMP and GMP. IMP and GMP are produced in Japan commercially by enzymatic hydrolysis of yeast RNA and by direct fermentation.

IMP production by fermentation involves either microbial production of inosine, which is chemically phosphorylated to IMP, or direct fermentation to

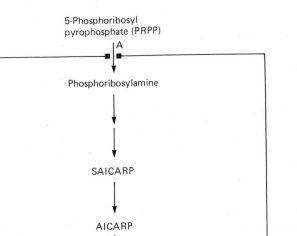

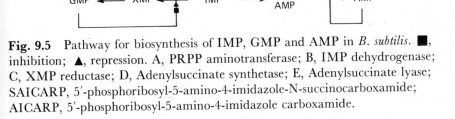

**Fig. 9.5** Pathway for biosynthesis of IMP, GMP and AMP in *B. subtilis*. ■, inhibition; ▲, repression. A, PRPP aminotransferase; B, IMP dehydrogenase; C, XMP reductase; D, Adenylsuccinate synthetase; E, Adenylsuccinate lyase; SAICARP, 5′-phosphoribosyl-5-amino-4-imidazole-N-succinocarboxamide; AICARP, 5′-phosphoribosyl-5-amino-4-imidazole carboxamide.

IMP. The normal cell membrane is permeable to inosine but not to IMP.

The pathway for production of purine nucleotides by *Bacillus subtilis* is illustrated in Fig. 9.5. In a sequence of metabolic reactions 5-phosphoribosyl pyrophosphate (PRPP) is converted to IMP which is a branch-point precursor of GMP and AMP. The specific activity of IMP dehydrogenase is much higher than that of adenylsuccinate synthetase, resulting in the predominant conversion of IMP to GMP. XMP and GMP cause feedback inhibition and repression of IMP dehydrogenase. Adenylsuccinate synthetase and adenylsuccinate lyase are regulated by AMP. AMP and GMP to a lesser extent cause feedback inhibition of PRPP aminotransferase.

When the cell synthesizes IMP it can be dephosphorylated by the cell and excreted into the medium as inosine. Commercial inosine-producing strains are deregulated adenine and guanine auxotrophs. Typical IMP producers are adenine auxotrophs having low activity of IMP-degrading enzymes. They are permeability mutants capable of excreting IMP. Mutants of *Brevibacterium ammoniagenes*, used for industrial production, are also insensitive to $Mn^{2+}$ which

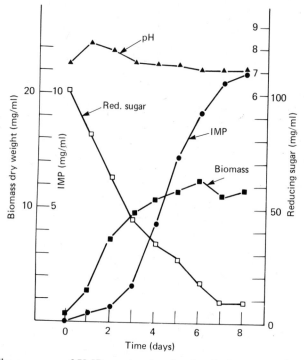

**Fig. 9.6** Time course of IMP production by *Brevibacterium ammoniagenes* Ky 13102 (reproduced with permission from Furuya *et al.*, 1968).

normally causes decreased IMP production. The pattern of production of IMP by *B. ammoniagenes* is illustrated in Fig. 9.6.

The major problem with over-production of GMP by direct fermentation relates to the feedback regulation of PRPP aminotransferase by GMP. GMP may be produced by a number of fermentation procedures:

(1) Production of guanosine by direct fermentation, using adenine auxotrophs to enhance production of GMP from glucose. GMP does not permeate across the cell membranes and is excreted as guanosine which requires chemical phosphorylation to GMP.

(2) Use of a mixed culture of two mutants of *Brevibacterium ammoniagenes*, one capable of producing XMP from glucose and the other capable of converting XMP into GMP.

(3) Production of 5-amino-4-imidazolecarboxamide riboside (AICAR) by fermentation followed by its chemical conversion to GMP.

Yields of nucleosides or related products obtained by fermentation of mutant strains are illustrated in Table 9.3.

**Table 9.3** Production of nucleotides and related products by fermentation

| Microbial strain | | Product | Genetic characteristics | Yield $(g\ l^{-1})$ |
|---|---|---|---|---|
| *B. ammoniagenes* | Ky 13102 | IMP | Ade | 12.8 |
| | Ky 13105 | IMP | Ade⁻Mn²⁺ insensitive | 19.0 |
| | Ky 13714 | Inosine | Ade⁻Gua⁻6MG^r | 13.6 |
| | Ky 13761 | Inosine | Ade⁻Gua⁻6MG^r6MTP^r | 30.0 |
| *Microbacterium* sp. | NO 250 | Inosine | Ade⁻6MTP^r8AG^rMSO^r6TG^r | 35.0 |
| *B. subtilis* | C-30 | Inosine | Ade⁻His⁻Tyr⁻ | 10.5 |
| | RDA-16 | Inosine | Ade⁻His⁻Red⁻Dea⁻8AG⁻ | 18.0 |
| | MG1 | Guanosine | Ade⁻His⁻Red⁻MSO^rPsi^rDec^r | 16.0 |
| *B. ammoniagenes* | | GMP | Two mutants, one converting glucose to XMP, the other XMP to GMP | 2.5 |
| *B. megaterium* | 335 | AICAR | Mutant | 16.0 |

Resistance abbreviations: r, resistant; 8AG, 8-azaguanine; 6MG, 6-mercaptoguanine; 6MTP, 6-methylthiopurine; MSO, methionine sulfoxide; Psi, psicofuranine; 6TG, 6-thioguanine, Dec, decoyinine.
Auxotroph abbreviations: Ade, adenine; Dea, AMP deaminase; Gua, guanine; His, histidine; Red, GMP reductase; Tyr, tyrosine.

**Vitamins**

While micro-organism have the ability to synthesize a wide range of vitamins, they do not normally over-produce these compounds. Vitamin $B_{12}$ and riboflavin are produced commercially by fermentation although riboflavin is now mainly manufactured primarily by chemical synthesis.

VITAMIN $B_{12}$

Current annual production of vitamin $B_{12}$ is 10 000 kg. Production may be accomplished using a two-stage process involving strains of *Propionibacterium* or a single-stage process involving *Pseudomonas denitrificans*.

In the first *Propionibacterium* fermentation stage, 5'-deoxyadenosylcobinamide is mainly produced which is then converted to vitamin $B_{12}$ in the second fermentation stage. Both fermentation stages are aerobic and the overall yield of product obtained in a fermentation time of about 6 days is $40 \, \text{mg} \, l^{-1}$.

The single-stage *P. denitrificans* aerobic fermentation process is carried out in a medium supplemented with cobalt and 5,6-dimethylbenzimidazole. Betaine increases the product yield either by activating biosynthesis or increasing membrane permeability. Consequently beet molasses, which contains betaine, may be effectively used as carbon feedstock. Mutant strains of *P. denitrificans* produce around $60 \, \text{mg} \, l^{-1}$ product in a four-day fermentation.

RIBOFLAVIN

An aerobic fermentation process for production of riboflavin using *Ashbya gossypii* was introduced in 1947. At the end of a seven-day fermentation, the riboflavin, which is present both in solution and bound to the mycelium, is recovered in yields of $7–8 \, \text{g} \, l^{-1}$. Strong competition exists between chemical and microbiological processes and research effort is being directed towards the development of *Bacillus subtilis* strains which over-produce and excrete riboflavin. It is expected that the application of recombinant DNA techniques will result in the construction of *B. subtilis* strains capable of producing commercially-viable yields of riboflavin.

**Fats and oils**

It has frequently been argued that fermentation might provide a practical route for production of fats and oils, particularly in areas like Europe, where the vast majority of these commodities are imported. Based on the assumption that 5–6 tonnes of wheat would be required to produce 1 tonne of recovered lipid, it is estimated that use of an efficient fermentation route for lipid production would result in manufacturing costs of £2500–3000 per tonne of lipid. Since these costs are clearly not competitive with the prices of current refined vegetable oils (Table

**Table 9.4** Refined vegetable oil prices (1987). Reproduced with permission from *Enzyme and Microbial Technology*, 1987

| Vegetable oil | (£/tonne) |
|---|---|
| Commodity oils | |
| Groundnut | 540 |
| Soya | 270 |
| Colza/rape | 880 |
| Sunflower | 300 |
| Cotton | 400 |
| Copra | 270 |
| Palm kernel | 220 |
| Linseed | 500 |
| Corn | 350 |
| Palm | 230 |
| Olive | 1 170 |
| Specialty oils | |
| Castor | 1 170 |
| Cocoa butter | 3 000 |
| Jojoba | 10 000 |
| Evening primrose | 35 000 |

9.4) industrial fermentation routes for production of commodity lipids from carbohydrates do not represent a viable option and opportunities only exist for using microbial technology to produce specialty oils.

## γ-LINOLENIC ACID PRODUCTION

γ-Linolenic acid is currently used as a dietary supplement and as a precursor in prostaglandin synthesis. Evening primrose oil which contains up to 7% γ-linolenic acid represents a plant source of this lipid. Some moulds from the order Mucorales, *Cunninghamella elegans* and *Mortierella isabellia*, produce 3–5% of their biomass as linolenic acid. Recently John E. Sturge (UK) announced that it had started up commercial production of linolenic acid by fermentation. The fermentation, involving a *Mucor* species, is carried out in stirred $220\,m^3$ fermenters, and uses pure glucose as major substrate. The refined oil is recovered by solvent extraction of the cells.

# Chapter 10

# *Health care products*

## Antibiotics

Antibiotics are secondary metabolites which inhibit the growth processes of other organisms. The observation of Alexander Fleming in 1929 that staphylococcal growth was inhibited by *Penicillium notatum* led to the development of production processes for penicillin and started the antibiotic era. Currently more than 6000 substances having antibiotic activity are known. About 100 antibiotic types are produced by industrial fermentation while nearly 50 semi-synthetic compounds also have clinical applications as antibiotics. Annual worldwide antibiotic production exceeds 100 000 t with an estimated market value of $5 billion. The most widely marketed antibiotics are the $\beta$-lactams (penicillins and cephalosporins) and the tetracyclines. The microbial antibiotics, chloramphenicol and pyrrolnitrin are now manufactured using cheaper chemical methods.

Antibiotics are used primarily as antimicrobial agents in human disease/infection therapy. Other applications include their use as cytotoxic agents against certain tumour types, as disease-control agents in veterinary medicine and plant pathology, as food preservatives and as animal growth promoters. Selected examples of important antibiotics produced by fermentation for pharmaceutical use are given in Fig. 10.1. Many of the examples given describe one member of a family of related antibiotics.

PENICILLINS

The basic structure of the penicillins is 6-aminopenicillanic acid (6-APA) made

## 10.1 Examples of important antibiotic types produced by fermentation for pharmaceutical use

| Antibiotic type | Example | Producing organism | Activity spectrum[1] | Structure[2] | Other comments |
|---|---|---|---|---|---|
| β-lactam | Penicillin G | *Penicillium thrysogenum* | G⁺ | | Low toxicity, acid labile, β-lactamase sensitive |
| | Ampicillin | *Penicillium chrysogenum* | G⁺G⁻ | | β-Lactamase sensitive, acid-stable |
| | Cephalosporin C | *Cephalosporium acremonium* | G⁺G⁻ | | Low toxicity, penicillinase-resistant but inactivated by β-lactamases produced by some G⁻ bacteria |
| Peptide | Bacitracin | *Bacillus licheniformis* | G⁺G⁻ | | Use confined to topical application because of toxicity |
| Aminoglycoside | Streptomycin | *Streptomyces griseus* | G⁻ | | Mainly used to treat tuberculosis |
| Macrolide | Erythromycin | *Streptomyces griseus* | G⁺ | | Particularly effective against *Staphylococci* and diphtheroids. Low toxicity |
| Polyene macrolide | Candidin | *Streptomyces viridoflavus* | F | | Widely used for topical anti-fungal application |
| Tetracycline | Chlortetracycline | *Streptomyces aureofaciens* | G⁺G⁻ | | |
| Aromatic | Griseofulvin | *Penicillium patulum* | F | | |

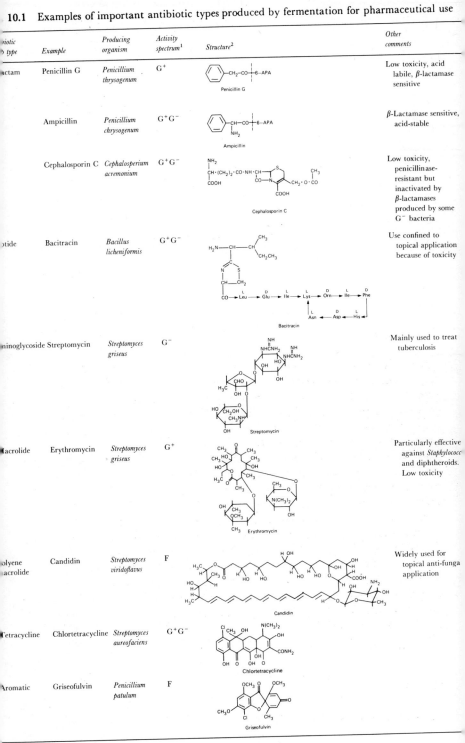

1. G⁺, Gram-positive; G⁻, Gram-negative; F, fungi.
2. 6-APA = 6-aminopenicillanic acid.

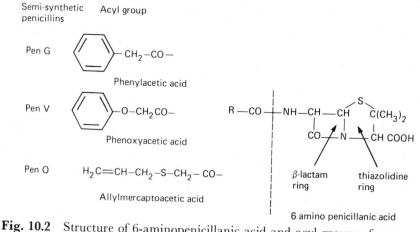

**Fig. 10.2**   Structure of 6-aminopenicillanic acid and acyl groups of some semi-synthetic penicillins.

up of a thiazolidine ring fused with a $\beta$-lactam ring (Fig. 10.2). The 6-amino position carries a variety of acyl substituents. In the absence of addition of side-chain precursors to the fermentation medium a mixture of natural penicillins is produced but only benzylpenicillin (Pen G) and phenoxymethylpenicillin (Pen V) are therapeutically important. Both have a similar target spectrum (Gram-positive bacteria) but Pen G is acid-labile and must be administered parenterally whereas the acid stable Pen V may be taken orally. While Pen G is produced naturally, better control of the fermentation process is achieved, resulting in higher yields of Pen G requiring simpler down-stream processing, by addition of the phenylacetic acid precursor to the medium. By incorporation of precursors phenoxyacetic acid and allylmercaptoacetic acid into the medium, phenoxyme-thylpenicillin (Pen V) and the less allergenic allylmercaptomethylpenicillin (Pen O) may be produced. Penicillin derivatives, having improved stability and antimicrobial activity, may be produced by semi-synthetic processes, following chemical or enzymatic hydrolysis of Pen G to 6-APA.

*Penicillin biosynthetic pathway*

The $\beta$-lactam thiazolidine is synthesized from L-$\alpha$-aminoadipate, L-cystine and L-valine by formation and cyclization of a peptide to produce isopenicillanic acid. Benzylpenicillin is then produced by transacetylation. Lysine and penicillin share a common anabolic pathway to L-$\alpha$-aminoadipic acid and lysine is an inhibitor of penicillin synthesis. The metabolic pathway and some of the possible regulatory mechanisms involved in benzylpenicillin production are illustrated in Fig. 10.3. In addition, glucose causes catabolite repression of penicillin biosynthesis and penicillin appears to regulate its own synthesis. Penicillin production is also affected by phosphate concentration.

*Strain development*

Initial yields obtained with Fleming's *P. notatum* strain, were 2 International Units/ml or 1.2 mg/l. Isolation in 1943 of *P. chrysogenum* NRRL-1951, a strain

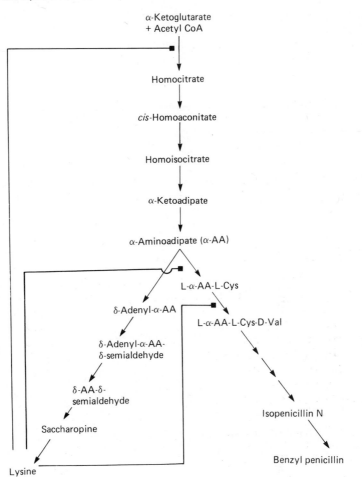

**Fig. 10.3** Outline pathway of penicillin biosynthesis indicating possible points of regulation by lysine.

which was more suitable for submerged culture than the original *P. notatum* strain, raised the yield to 120 IU/ml. Mutation of this strain produced the famous Wisconsin strain Wis Q 176, which yielded 900 IU/ml. Mutation/selection techniques involving X-rays, short-wave UV radiation and chemical mutagens, such as methylbis-($\beta$-chloroethyl)amine, nitrosoguanidine, alkylating agents and nitrite, were used until the early 1970s. Studies with blocked mutants led to an understanding of the biosynthetic pathways which in turn suggested appropriate selection techniques. The discovery of the parasexual cycle in *P. chrysogenum* led to strain improvement using parasexual breeding and protoplast fusion techniques. Through continued strain-improvement programmes, combined with fermentation optimization, typical current penicillin yields of 85 000 IU/ml or 50 g/l have been achieved.

*Production of penicillin*

Optimized inoculum spore concentration and formation of pellets in loose rather than compact form in the vegetative growth stages is essential for attainment of high penicillin yields. Biomass doubling time is usually about 6 h. Strain stability problems exist and careful strain maintenance is required. Typical penicillin production conditions involve use of a fed-batch culture fermentation with a medium containing corn steep, ammonia, salts and a carbon source such as glucose, lactose or molasses. Most fermentation processes included corn steep as organic nitrogen source because it improved penicillin yield due to its content of side-chain precursors. Introduction to the medium of specific side-chain precursors enabled other nitrogen sources to replace corn steep. The continuous maintenance of ammonia in the medium supports respiration, prevents mycelial lysis and is important for penicillin synthesis. pH is controlled at 6.5 and phenylacetic acid or phenoxyacetic acid is continuously fed as precursor. The rate of sugar utilization and the rate of oxygen supply are important fermentation parameters. The oxygen supply rate is especially critical because the increasing viscosity of the broth hinders oxygen transfer. The process requires an oxygen uptake rate of 0.4–1.0 mmol per litre per minute and an RQ (moles $CO_2$ formed/moles $O_2$ consumed) of about 0.95. Figure 10.4 illustrates an example of a fermentation outline including the patterns of substrate utilization and product formation. The industrial fermentation is typically characterized by a high growth rate for about two days. Then growth rate declines and the rate of penicillin formation increases and continues production for a further period of 6–8 days provided the appropriate substrate feeds are maintained.

6-APA is produced from Pen G by passage through an immobilized penicillin acylase column which converts Pen G to phenylacetic acid and 6-APA. The column pH is maintained at neutrality by NaOH addition and the 6-APA in the column eluent is recovered by precipitation at pH 4.0. In production of semi-synthetic penicillins, the resultant 6-APA is chemically acylated with the appropriate side-chain using standard methods.

## TETRACYCLINES

The basic structure of tetracycline consists of a naphthacene ring (see Fig. 10.1). Clinically important tetracyclines, produced by fermentation or semi-synthesis, vary with respect to specific ring substituents. Chlortetracycline and oxytetracycline are the major tetracyclines produced by *Streptomyces* whereas tetracycline is normally only formed in minor amounts. *Streptomyces aureofaciens* strains, mutated to block the chlorination reaction, excrete tetracycline as the major product.

*Tetracycline biosynthetic pathway*

Chlortetracycline synthesis is a complex metabolic pathway having 72 intermediates and involving more than 300 genes. Initial stages involve formation of malonamoyl CoA bound to the enzyme complex anthracene synthase. Malonamoyl CoA condenses with 8 molecules of malonyl CoA and cyclization occurs

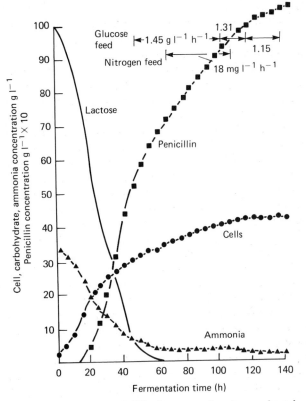

**Fig. 10.4**  Time course of a penicillin fermentation (reproduced with permission from Queener and Swartz, 1979).

with eventual formation of chlortetracycline. High-yielding tetracycline strains are characterized by a lower rate of glycolysis and chlortetracycline production may be enhanced by use of the glycolysis inhibitor benzylthiocyanate. Under these conditions activity of the pentose phosphate cycle increases. The rate-limiting enzyme in chlortetracycline biosynthesis may be anhydrotetracycline oxygenase, the second last enzyme in the biosynthetic pathway. Its activity appears to be proportional to the rate of antibiotic synthesis. Synthesis of this enzyme is repressed by phosphate and stimulated by benzylthiocyanate. There is also an inverse relationship between the level of adenylates and activity of this enzyme. The ATP level or the total adenylate level appears to act as the metabolic effectors in catabolite regulation of tetracycline biosynthesis.

*Production of chlortetracycline*
Current industrial yields of tetracyclines are around $20\,000\,\mu g\,ml^{-1}$. Because of the complexity of the biosynthetic pathway, strain yield improvement has depended solely on mutation/selection techniques. Selection of strains resistant to the produced antibiotic is another method which has been applied to improving

production capacity. Typical fermentation production media contain sucrose, corn steep, ammonium phosphate and salts with pH and temperature maintained at 5.8–6.0 and 28°C, respectively. High aeration rates are necessary particularly in the biomass growth stages. If glucose is used, continuous feeding is necessary. Because of phosphate repression, tetracycline fermentations are run under phosphate-limited conditions.

Production of chlortetracycline in submerged culture may be subdivided into three phases. The first phase is characterized by a rapid increase in biomass and rapid consumption of nutrients. During this phase the mycelium is characterized by the presence of thick basophilic hyphae with a high RNA content. In the second phase, growth rate decreases and sometimes ceases, maximum rates of antibiotic synthesis are observed and the organism differentiates. Hyphal filaments appear thin and contain a low RNA content. In the third phase, lower rates of antibiotic production are observed and mycelium fragmentation and lysis occurs.

*Streptomyces aureofaciens* produces a certain proportion of tetracycline in addition to chlortetracycline. Chloride ions are necessary for chlortetracycline formation, particularly in high producing mutants. Other agents including fluoride ions, copper, methionine and 5-fluorouracil suppress production of tetracycline.

A low content of chloride ions in the medium is the basic condition for production of tetracycline by *S. aureofaciens*. Oxytetracycline is produced by *Streptomyces rimosus* under phosphate-limited conditions.

## CURRENT ANTIBIOTIC FERMENTATION RESEARCH

The novel $\beta$-lactam thienamycin, produced by *Streptomyces cattleya* and discovered in 1976, has an unusual and highly-desirable antibiotic spectrum against Gram-positive and Gram-negative bacteria. It is not attacked by $\beta$-lactamases produced by strains resistant to many penicillins and cephalosporins.

Because of the large number of genes involved in the biosynthesis of an antibiotic, recombinant DNA research aimed at antibiotic strain development is extremely complex. A major objective of applying modern DNA techniques to antibiotic strain development will be to increase the rate of the rate-limiting biosynthetic step(s), possibly by increased gene dosage. However account must also be taken of possible transcriptional, translational or substrate level regulation as well as product secretion processes. Recombinant DNA technology is now being applied to *Aspergillus*, *Streptomyces*, *Penicillin* and *Cephalosporium* species and is expected to quickly lead to manipulation of pathways of secondary metabolites via gene isolation, sequencing, amplification and *in vitro* mutagenesis. Knowledge of the genetic loci controlling the biosynthetic pathway is of the utmost importance for the exploitation of DNA techniques of protoplast fusion and gene cloning with the aid of plasmid vectors.

A transformation system for *Cephalosporium* has been developed and the gene for *C. acremonium* cyclase has been cloned. Cloning of the remaining genes of *C. acremonium* and of the *P. chrysogenum* $\beta$-lactam biosynthetic pathway may allow substantial de-regulation of the terminal steps. These organisms may also prove

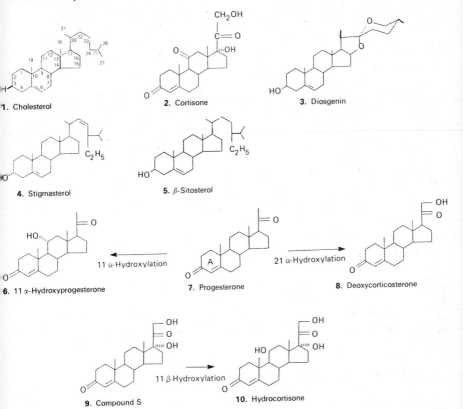

**Fig. 10.5** Chemical structures of some sterol compounds and examples of key biotransformation reactions.

attractive as hosts for introduction of genes from *Streptomyces* species that produce β-lactam antibiotics. β-Lactam antibiotics, having better oral absorption properties and a broader activity spectrum are expected to dominate the antibiotic industry in the future.

## Steroid fermentations

A variety of important steroid compounds is produced by routes which require one or more biosynthetic steps. Chemical structures of a few of these compounds and the reactions catalysed are given in Fig. 10.5. They are used for treatment of a variety of diseases and injuries. In the 1940s the naturally-occurring corticosteroids, cortisone (**2**) and cortisol (hydrocortisone) (**10**) were successfully used in the treatment of inflammatory and allergic diseases. Various plant and animal compounds were examined as potential starting materials for synthesis of this new class of compounds. Cholesterol (**1**) was chemically unsuitable and it soon became evident that diosgenin (**3**) (barbasco plant root) and stigmasterol (**4**) (soybean seed oil) were possible potential new materials for progesterone and pregnenolone synthesis, respectively.

Pregnenolone is easily converted, chemically, to progesterone (**7**) which has the proper A-ring configuration of corticosteroids. Introduction of a 21-hydroxyl into progesterone gives deoxycorticosterone (**8**), a naturally-occurring corticosteroid. Production of cortisone and hydrocortisone still requires introduction of an 17α-hydroxyl group and an 11-hydroxy or 11-keto group. The 17α-hydroxyl group could be introduced chemically and microbiologically and superior chemical routes are currently used. A major advance was made by Murray and Peterson in 1952 who discovered a process for bioconversion of progesterone to 11α-hydroxyprogesterone (**6**), involving *Rhizopus arrhizus*. Later, a wide range of bioconversion reactions specific for various positions of the steroid molecule were characterized. The steroid biotransformations of greatest practical importance are the 11-hydroxylations, 16α-hydroxylations and 1-dehydrogenation.

## BIOTRANSFORMATION PROCESSES

In steroid biotransformation processes, the organisms containing the required enzyme(s) are produced by conventional submerged fermentation processes under conditions which induce synthesis of the enzyme. When growth and enzyme production has occurred, the water-insoluble steroid substrate is added, as a powder slurry or dissolved in an organic solvent, and the bioconversion takes place. After the bioconversion is completed, particularly in more highly-developed fermentations conducted at high steroid substrate levels, most of the product is recovered by filtration with the biomass fraction and is separated by extraction using organic solvents.

*Rhizopus nigricans* and *Aspergillus ochraceus* are used commercially for 11α-hydroxylation of steroids. The α-hydroxylase is induced in the fermentation by progesterone. An undesirable 6β-hydroxylase is also induced and fermentation conditions are designed to minimize production of this enzyme.

The most commonly used micro-organisms for steroid 11-β-hydroxylation are *Curvularia lunata* and *Cunninghamella blakesleeana*. 17α,21-Dihydroxypregn-4-ene-3,20-dione (compound S) (**9**) is the substrate used to produce hydrocortisone (**10**) and again other by-products are formed. Compound S and other substrates induce the enzyme.

Many micro-organisms can catalyse the 1-dehydrogenation reaction of steroids. Hydrocortisone is a typical substrate for production of prednisolone. In each organism investigated the 1-dehydrogenase is inducible, but in the case of one well-studied organism, *Septomyxa* sp., cortisone and hydrocortisone do not induce while progesterone and BNA (3-ketobisnor-4-cholen-22-al) are good inducers. Induction occurs only after glucose exhaustion.

## TRENDS IN STEROID TRANSFORMATION RESEARCH

Recent research on the microbial degradation of side-chains of abundantly-available sterols such as cholesterol (from wool) and $\beta$-sitosterol (5) (from soybean oil) to produce useful steroids is receiving much attention. Degradation of the steroid rings appears in many instances to occur more rapidly than cleavage of the sterol side-chain and methods are being developed which inhibit this ring degradation.

Chemical addition of a 9$\alpha$-fluorine atom to some corticosteroids greatly improves their anti-inflammatory activity but also undesirably increases salt retention in humans. This side-effect could be diminished by microbial 16$\alpha$-hydroxylation. Conversions of about 50% could be achieved using *S. argenteolus* with formation of 2$\beta$-hydroxy by-products which can be minimized by proper mutant selection.

Traditionally, steroid bioconversions have been carried out using conventional fermentation technology followed by the biotransformation reaction. Use of acetone-dried cells of *Arthrobacter simplex* for 1-dehydrogenation of 9$\alpha$-fluoro-16$\alpha$-hydroxycortisone greatly decreases the level of 20-ketoreductase and undesirable by-product formation. Addition to the reaction mixture of an artificial electron acceptor, menadione, was required. Hydrocortisone was converted to prednisolone when the substrate, dissolved in ethanol and phenazine methosulphate, was passed through an immobilized 1-dehydrogenase enzyme column. Use of cells or enzymes in organic solvents, which solubilize the steroid substrate and products, offers the possibility to use much higher substrate concentrations in bioconversions. Use of whole cells of *Corynebacterium simplex* instead of isolated immobilized enzymes may eliminate the need for addition of an electron acceptor and reduces the amount of enzyme losses during immobilization.

Recovered resuspended spores of *Septomyxa affinis*, obtained by growth on solid or in liquid media, can 1-dehydrogenate various types of steroids. *Aspergillus ochraceus* spores have been used to convert progesterone to 11$\alpha$-hydroxyprogesterone. *Rhizopus nigricans*, immobilized in alginate or agar gels, but not in polyacrylamide gels, could $\alpha$-hydroxylate progesterone. 11$\beta$-Hydroxylation of compound S by *C. lunata* immobilized in cross-linked polyacrylamide gels has also been demonstrated.

## Ergot alkaloids

Ergot alkaloids have a wide variety of therapeutic uses including treatment of migraine and other vascular headaches, uterine atonia, circulatory disturbances, hypertension and Parkinsonism. There are currently over 40 known ergot alkaloids produced by various strains of the parasitic ascomycete, *Claviceps*. The structure consists of *d*-lysergic acid (or its stereoisomer *d*-isolysergic acid) linked with a tricyclic peptide or an amino alcohol by an amide bond. The structures of some naturally-occurring ergot alkaloids are illustrated in Fig. 10.6.

Total chemical synthesis of ergot alkaloids is possible but currently is not cost-effective. Traditionally, ergot alkaloids were obtained by mechanical infection of rye flowers with *Claviceps* and subsequent harvesting and extraction of sclerotia. The process is therefore seasonal and weather-dependent.

Three species of *Claviceps* are used for alkaloid production by fermentation,

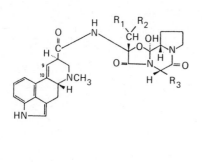

| $R_1$ | $R_2$ | $R_3$ | Name |
|---|---|---|---|
| H | H | $CH_2-\bigcirc$ | Ergotamine |
| H | H | $CH_2CH(CH_3)_2$ | Ergosine |
| $CH_3$ | $CH_3$ | $CH_2-\bigcirc$ | Ergocristine |
| $CH_3$ | $CH_3$ | $CH_2CH(CH_3)_2$ | $\alpha$-Ergocryptine |
| $CH_3$ | $CH_3$ | $CH(CH_3)CH_2CH_3$ | $\beta$-Ergocryptine |
| $CH_3$ | $CH_3$ | $CH(CH_3)_2$ | Ergocornine |
| H | $CH_3$ | $CH_2-\bigcirc$ | Ergostine |

**Fig. 10.6**  Structures of naturally-occurring ergot alkaloids.

namely *C. paspali*, *C. fusiformis* and *C. purpurea*. Whereas the original strain of *C. paspali* produced only $20\,\mu g\,l^{-1}$ alkaloid, strain development and medium optimization raised yields to greater than $5\,g\,l^{-1}$. Alkaloid composition is determined by a combination of the strain and culture conditions used. High rates of alkaloid formation appear to be linked to the organism's ability to simultaneously metabolize high sucrose and citrate concentrations in phosphate-deficient media. Synthesis proceeds in parallel with lipid and sterol anabolism. Production strains are shear-sensitive, have a high oxygen requirement and are unstable.

## Microbial recombinant DNA products

*Escherichia coli* has been the organism of choice as a host for rDNA product formation owing to the well-characterized genetic systems of this organism. This has resulted in the achievement of high levels of expression. The organism is also easily grown at high growth rates on defined media to give high cell densities.

The first therapeutic agent produced by rDNA technology to achieve regulatory approval was human insulin. This hormone is produced in fermentation using a recombinant strain of *E. coli*. According to Eli Lilly & Co., who manufacture this hormone by fermentation, the product's chemical and physical structures are indistinguishable from natural human insulin. Novo Industri developed alternative technologies for commercial production of human insulin. Initially their process involved conversion of porcine insulin to human insulin and later human insulin was produced by fermentation using a recombinant yeast strain. Human growth hormone produced by recombinant *E. coli* is being marketed for treatment of hypopituitary dwarfism. Interleukin-2 (IL-2) or lymphocyte growth factor is another rDNA product produced by *E. coli*. IL-2 has been reported to stimulate the body's T-cells to attack and diminish the size of discrete tumours. A fermentation profile for production of recombinant IL-2 is illustrated in Fig. 10.7.

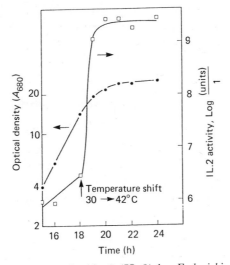

**Fig. 10.7** Production of interleukin-2 (IL-2) by *Escherichia coli* K-12 containing a plasmid coding for IL-2 under control of a temperature-sensitive promoter. At 30°C bacterial propagation is promoted but IL-2 synthesis is repressed. Rapid IL-2 synthesis is induced by a temperature shift to 42°C (reproduction with permission from Bauer. See Khosrovi and Gray, 1985).

Interferons are proteins synthesized by most cells of higher animals in response to viral infections. Their production by infected cells serves to 'interfere' with the spread of the infection to healthy cells. Interferons also act as antitumour agents for certain types of cancer. The severe shortage of purified human interferons has hampered clinical trials of their therapeutic value. α-Interferon, successfully cloned into *E. coli*, was produced commercially and has been used in a wide range of clinical trials. Interferons produced by bacteria in this way lack the glycoprotein component of native interferons. A variety of other compounds have been cloned into *E. coli*, including β- and γ-human interferons, various animal interferons, various human and animal growth hormones, epidermal and other growth factors, lymphokines, human serum albumin, plasminogen activators and a variety of enzymes.

*Escherichia coli* is disadvantageous as a host in that recombinant proteins are generally not secreted and cells may contain endotoxins and pyrogenic lipopolysaccharides which must be removed during purification. *Bacillus subtilis* is another host which has been used for production of recombinant products. It is non-pathogenic, grows under aerobic conditions, does not produce lipopolysaccharides and secretes extracellular proteins directly into the medium. Where genes have been cloned from one *Bacillus* species into another high expression levels have been observed. However, unlike *E. coli*, expression levels have been reported to fall by several orders of magnitude when heterologous genes are cloned into *B. subtilis*. Yeasts, because of their long-time use in food fermentations, are accepted as safe organisms, lack pyrogenic lipopolysaccharides and can secrete and glycosylate recombinant proteins. Expression levels quoted for proteins cloned into yeast are lower than those quoted for *E. coli*. *Saccharomyces cerevisiae* has been engineered to produce commercial products such as hepatitis B surface antigen vaccine, approved by the US Food and Drug Administration, superoxide dismutase, epidermal growth factor and variety of other products. Secretion systems have been successfully developed for production of some recombinant proteins.

Filamentous fungi, such as *Aspergillus niger*, can produce up to 20 g/l of enzyme from a single glucoamylase gene copy. They can be engineered to express and secrete heterologous proteins of bacterial and mammalian origin and basic processes of secretion in fungi and mammalian cells are very similar. Fungal systems are potentially modifiable to more closely approximate mammalian cell processing of secreted proteins although a greater understanding of fungal glycosylation pathways is required. These organisms offer much promise as hosts for synthesis of large quantities of heterologous mammalian proteins.

## Vaccines

A number of vaccine types exist: live or killed microbial cells or viruses, natural or modified extracellular products such as toxins and subcellular fractions of cells and viruses. In the case of live vaccines, the organisms are able to replicate in the recipient. The efficacy of *Mycobacterium tuberculosis* vaccine, for example, depends on a certain level of multiplication in the human body. Live poliovirus vaccine is a second example. These live vaccines have been rendered harmless by attenuation, a treatment which essentially eliminates the disease-causing ability of the organism. Alternatively, live vaccines may be produced by bacteria or viruses closely related to the disease-causing organism which do not cause disease but can induce immunity, for example, smallpox vaccine. Killed vaccines are whole cells or cell components, generally derived from the virulent disease-bearing agent, which have been inactivated by chemical treatment. Extracellular microbial products, such as toxins, can be used as vaccines. They are first de-toxified by treatment with chemical reagents which destroy their toxic properties while leaving the immunogenic activity intact.

### VACCINE PRODUCTION

*Bacterial cells*

Most bacterial vaccines are produced in batch submerged culture processes. *Bordetella pertussis* (whooping cough) bacteria are grown in media containing ingredients such as acid-hydrolyzed casein, minerals and growth factors. The time course of a fermentation process is illustrated in Fig. 10.8. The risk of exposure of operators to the cells can be reduced by harvesting the cells by acid precipitation. When the cells are acidified to pH 4.0 they quickly sediment and can be collected in 3–5% of the culture volume. Cells are killed and de-toxified either by heating, by addition of sodium ethylmercurithiosalicylate or by a combination of these

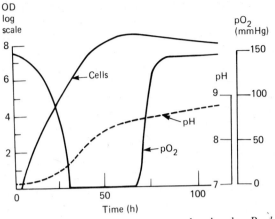

**Fig. 10.8** Fermentation profile of biomass production by *Bordetella pertussis*. Culture volume, 7 l; aeration 0.2/vvm; stirrer speed, 450 rpm (reproduced with permission from Van Hemert, 1974).

treatments, to produce the pertussis vaccine. *Salmonella typhi* and *Vibrio cholera*, used as vaccines against typhoid and cholera respectively, are produced in media containing complex nitrogen with glucose added during culture. Control of pH and $pO_2$ in the fermentation gives rapid cell growth and high cell yields. Fermentation profiles of *V. cholerae* and *S. typhi* are illustrated in Figs 10.9 and 10.10, respectively. A recovery/inactivation system for *S. typhi* involves suspension of cells in acetone for 24 h at 37°C followed by cell lyophilization. Other examples of inactivated bacteria are the cholera and plague vaccines containing inactivated strains of *Vibrio cholera* and *Yersinia pestis*, respectively. *Mycobacterium tuberculosis* (BCG vaccine) grows as a coherent pellicle on the surface of fluid media and was conventionally grown in glass containers on a shallow layer of medium. The

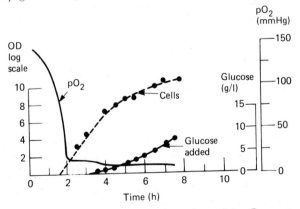

**Fig. 10.9** Profile of a *Vibrio cholerae* fermentation with $pO_2$ control at 15 mm Hg and pH control at 7.3 (reproduced with permission from Van Hemert, 1974).

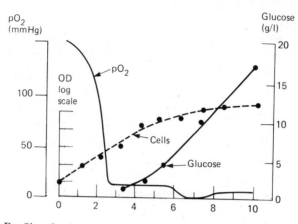

**Fig. 10.10** Profile of a fermentation for production of *Salmonella typhi*. Fermentation volume, 7 litres, gas-flow rate, 0.43 vvm, stirrer speed, 450 rpm; pH controlled at 7.6 with NaOH (reproduced with permission from Van Hemert, 1974).

biomass is compressed into a dense cake and a cell suspension prepared by shaking the pellicle with steel balls. A submerged culture process became the method of choice for preparing material for lyophilization. The cells are dispersed in a complex medium using Tween 80. A typical fermentation profile is illustrated in Fig. 10.11. Because viable cells are required for the vaccine, the fermentation recovery and cell storage methods are designed to maintain maximum cell viability.

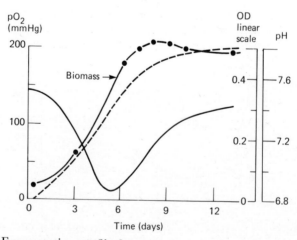

**Fig. 10.11** Fermentation profile for production of BCG cells (reproduced with permission from Van Hemert, 1974).

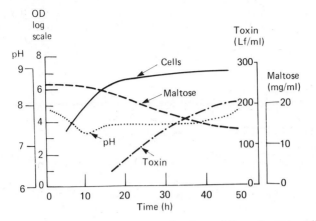

**Fig. 10.12**  Fermentation profile for production of *Corynebacterium diphtheriae* toxin (reproduced with permission from Van Hemert, 1974).

*Bacterial toxins*
The diphtheria toxoid vaccine is produced and secreted by *Corynebacterium diphtheriae* in a complex medium in submerged culture (Fig. 10.12). Iron concentration is important for toxin production as excess iron inhibits synthesis. The toxin may contribute up to 75% of all extracellular protein secreted. *Clostridium tetani* is an obligate anaerobe which produces the tetanus toxin. Its ability to form thermostable spores led the WHO to set down specific safety precautions governing its production. A production profile of the toxin, indicating the relative concentrations of intracellular and extracellular toxin during fermentation, is illustrated in Fig. 10.13. Detoxification is accomplished by use of formaldehyde under controlled reaction conditions. Crude toxoid preparations contain many impurities and antigens are usually further purified. A widely-used method involves sequential fractionation of the material with methanol at acid pH, followed by lyophilization of the product. Other purification methods used include dialysis, gel filtration and DEAE-cellulose chromatography.

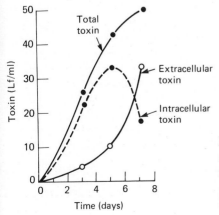

**Fig. 10.13**  Production of intracellular and extracellular toxin during fermentation of *Clostridium tetani* (reproduced with permission from Van Hemert, 1974).

*Viral vaccines*

Production of viral vaccines for human application, whether live or killed, is usually accomplished using *in vitro* animal-cell culture. Primary cells (i.e. cells taken directly from normal tissue and sub-cultured only once) or diploid cell lines have been used rather than aneuploid cell lines because of supposed risks of oncogenicity. Established cell lines are, however, used for some veterinary vaccines. The timing of the viral propagation step depends on whether the host cells die or continue to replicate after infection. Technology involving viral proliferation in hens' embryonated eggs is still used, however, for production of some vaccines.

Live poliovirus vaccine is a mixture of three types of attenuated polioviruses which have been propagated in cultures of monkey kidney cells or human diploid cell lines. In inactivated poliomyelitis vaccines, the viruses are treated with formaldehyde. Inactivated rabies virus is produced by growth on human diploid cells. Live attenuated viral vaccines are used for immunization against rubella, measles, mumps and yellow fever. Rubella virus is propagated in human diploid cells whereas primary chick-embryo cells are used to support proliferation of measles, mumps, yellow fever and smallpox.

Two types of inactivated influenza viruses are produced from viruses propagated on embryonated chick eggs. Whole killed virus is inactivated by treatment with formaldehyde or $\beta$-propiolactone. Split virus is prepared by treatment with agents such as cetyltrimethylammonium bromide (CETAB), sodium dodecylsulphate and Triton X-100, butyl and ethyl acetates, and Tween 80 and tri-*n*-butyl phosphate.

*Isolated antigen vaccines/subunit vaccines*

Whole-cell vaccines often contain toxic components, which cannot be easily inactivated, in addition to the desired immunogenic antigens. Since whole cells contain the complete genetic material of the pathogen, if the pathogen is not completely killed or attenuated, the vaccine itself may be capable of causing disease. There is also the possibility that attenuated strains might revert to virulence. Another problem with some vaccines is that they sometimes do not immunize the recipient against all strains of the pathogen. Some are also unstable in storage.

Attempts have been made to isolate subcellular fractions which retain immunogenicity but lack toxicity. For example, an acellular toxin vaccine, used in Japan, contains the key antigen components filamentous hemagglutinin (FHA) and lymphocytosis promoting factor (LPF). Toxicity associated with LPF is inactivated with formalin. The preparation has a greatly reduced content of lipopolysaccharide endotoxin. Pneumonia vaccines consist of a mixture of purified capsular polysaccharides from 14 types of *Streptococcus pneumonia*. Purification is achieved by alcohol fractionation, centrifugation, treatment with cationic detergents (e.g. CETAB), proteases, nucleases or activated charcoal, ultrafiltration and lyophilization. Purified capsular polysaccharides of *Neisseria meningitidis*, serogroups A and C are used in meningitis vaccines.

*Vaccine development using* rDNA

Recombinant DNA technology is being used to develop subunit vaccines. Genes that encode for portions of the hepatitis B surface antigen have been cloned and the isolated surface antigens are effective as a vaccine. Approval to use the vaccine was given during 1986 and 1987. A vaccinia recombinant vaccine for immunization against smallpox has been developed which is both safe and cheap to produce. Other vaccines are being developed against viruses such as influenza, polio and herpes, using rDNA techniques. Protein antigens on the surface of the AIDS virus (HIV, human immunodepressant virus) have been isolated which induce antibodies in rodents. The demonstration that these antibodies kill the virus paved the way for the development of a rDNA-prepared vaccine for the virus. The first experimental vaccine against AIDS based on recombinant versions of the HIV virul antigen produced via insect cell culture was approved for clinical testing in 1987 by the US Food and Drug Administration.

The first genetically-engineered vaccine to be marketed was vaccine for pseudorabies, a herpes virus that infects swine, introduced in 1986. A variety of other rDNA veterinary vaccines are being developed. As our understanding of the molecular structures, which are responsible for immunogenicity, increases, potential exists for the development of safe, synthetic, vaccines (for example, short amino acid chains which mimic the relevant sites on a virus coat protein).

## Monoclonal antibodies

When the body is challenged with an invading substance, such as a virus or microbial cell, white blood cells, called B-lymphocytes, respond by producing hundreds of antibodies which selectively bind to the invading material. Immunization techniques have thus been used as a means of developing natural resistance to pathogenic organisms. Similarly antibodies, contained in antisera, have been widely used in medicine for disease diagnosis and also in therapy of selected human illnesses. Kohler and Milstein in 1975 developed the basic technology for production of monoclonal antibodies (MCAs), preparations of individual specific antibodies produced by cells derived from a single ancestor or clone. Monoclonal antibodies, having high binding specificity for a single site (epitope) on a molecule or cell-surface antigen, have the capability to differentiate between related antigens which may differ by only a single epitope.

MCAs are widely used in clinical analysis kits for detection of diseases and conditions ranging from pregnancy to cancer and are being tested in applications as disease or damage markers within the body (diagnostic imaging). Of even greater significance are the potential applications of MCAs in disease therapy. Examples of MCAs on the market include one product which blocks kidney rejection during transplants and another which scavenges any potentially dangerous excess of the drug digitalis, administered to patients with cardiovascular disease. Many applications also exist outside medicine. The 1987 market for MCAs was estimated at $300 million with more than 100 products on sale and is expected to rise to over \$1 billion by 1990. When MCAs are used in therapy, perhaps by the mid-1990s, the market is expected to soar to \$7 billion per annum.

## DEVELOPMENT OF HYBRIDOMA CELLS

Hybridomas are cells capable of MCA production and cell division, obtained by fusion of B-lymphocyte cells (capable of antibody production) with cancer cells (capable of cell reproduction). The myeloma fusion partner confers unlimited reproductive capacity *in vitro*.

A myeloma parental cell line which lacks the enzyme hypoxanthine phosphoribosyl transferase (HPRT) is incubated with spleen cells obtained following immunization with the desired antigen. Incorporation of polyethylene glycol in the incubation medium promotes fusion of the cells with resultant production of hybrid cells. The cells are separated after a few minutes, suspended in culture medium containing foetal calf serum and distributed into microtiter plate wells for incubation. Aminopterin and hypoxanthine are incorporated into the medium. The post-fusion medium (HAT) also contains thymidine, to provide pyrimidines via the salvage pathway.

Normal spleen cells cannot proliferate in culture. Similarly, because of HPRT deficiency, parental myeloma cells die in the presence of aminopterin, an antifolate agent which blocks the cells' alternative route for nucleotide synthesis. Consequently, only fused cells survive and develop into visible colonies. Culture supernatants of these hybridoma cells are assayed for the presence of antibody reacting with the desired immunizing antigen and selected cultures are assayed for antibody specificity. The hybridomas producing desired antibodies are cloned and stored in liquid nitrogen.

## PRODUCTION OF MONOCLONAL ANTIBODIES

A particular hybridoma clone may be injected into mice where it grows in the ascites fluid of the peritoneal cavity from which antibodies are readily collected. Antibody concentrations in excess of $10 \text{ mg ml}^{-1}$ may be produced in this way and one mouse can supply enough antibody for up to 20 000 diagnostic kits as only tiny quantities are required per test. Much larger quantities are required for therapeutic applications and consequently a shift towards *in vitro* production of MCAs by fermentation methods has been necessary. For example, one company, Charles River Biotechnical Services in the USA, produced over 3 kg of MCAs in mouse ascites fluid in 1985. Considering that production of 1 g MCA requires 200–500 mice, which are time-consuming and costly to care for, one can appreciate the scale of a 3 kg/year *in vivo* production facility.

Celltech (UK) appeared to have the world's largest MCA cell-culture facility in 1986, consisting of two 1000-litre fermenters, in addition to a 200-litre and several 100-litre reactors. A major disadvantage however, of conventional cell-culture fermentation systems is that mammalian cells cannot be maintained at densities greater than $2 \times 10^6 \text{ ml}^{-1}$. Under these conditions typical concentrations of MCA

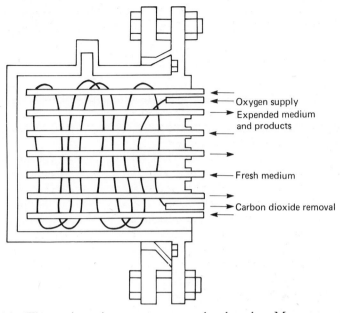

**Fig. 10.14** The static maintenance reactor developed at Monsanto (reproduced with permission from Van Brunt, 1986b).

produced are less than 75 $\mu$g ml$^{-1}$. Consequently there has been a substantial commitment to development of new mammalian cell-culture methods that increase both cell density and product concentration. A continuous culture system for MCA production developed by Monsanto Corporation, having a reactor volume of 16.5 litres is claimed to be equivalent in capacity to a 1000-litre conventional batch fermenter. Cells are first grown in a perfusion chemostat system, concentrated, mixed with a matrix material in a ratio of 1:10 and placed in a cylindrical vessel reactor. An array of porous tubes penetrates the reactor which circulate medium through the culture and a distributed semi-permeable membrane is used for $O_2$ supply and $CO_2$ removal (Fig. 10.14). Charles River extended its capability to *in vitro* large-scale production using the 'Opticel' system, in which the hybridoma cells are immobilized on a ceramic matrix. Bio-Response (USA) is able to produce up to 150 g MCA per day from an average mouse hybridoma culture growing in a 400-litre hollow-fibre reactor. Cell densities are maintained at less than $5 \times 10^7$ ml$^{-1}$. Advantages claimed for this system are that the cells appear to require lower serum concentrations than are needed for normal batch culture, which allows chemically-defined media with low serum content to be used. Damon Biotech (USA), has developed a specialized cell microencapsulation technology which involves immobilizing the hybridoma cells in hollow gelled sodium alginate spheres. Cells may grow and migrate within the microcapsule and capsule pore size is regulated to facilitate nutrient and metabolite diffusion while retaining the antibody. Microcapsules are cultured in 40-litre reactors equipped with variable-speed stirrers, $O_2$ and air/$CO_2$ supply

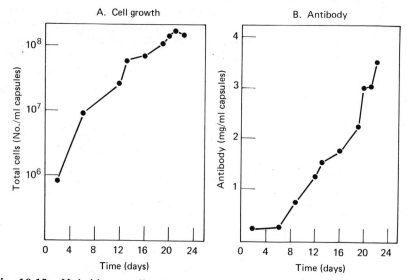

**Fig. 10.15** Hybridoma cell growth and monoclonal antibody production in microcapsules (reproduced with permission from Posillico, 1986).

lines and exhausts, and continuously supplied with medium throughout a 2–3 week culture period. During this period, cell and MCA concentrations within the microcapsules increase to greater than $10^8$ cells ml$^{-1}$ and 3 mg ml$^{-1}$, respectively (Fig. 10.15).

### DEVELOPING IMPROVED MONOCLONAL ANTIBODIES

Limited therapeutic trials with mouse MCAs have shown that they are effective at low dose rates and are very low in toxicity. However, mouse proteins are immunogenic in humans and patients quickly develop cross-reacting anti-mouse antibodies. While it is possible to make human hybridomas, the efficiency of transformation is very low and MCA production stability is variable. Methods are being developed which will enable human MCAs and hybrid antibodies (part human, part mouse-derived) to be expressed in mouse cell lines. Normal antibodies consist of four molecules, two identical heavy chains and two identical light chains. A second generation of antibodies, containing a single chain, is being developed. These are recombinant molecules, currently produced in *E. coli*, which consist of portions of the variable regions of an antibody's heavy and light chains covalently linked into one molecule by a peptide spacer (Fig. 10.16). Claimed advantages of these antibodies are their smaller size, greater stability and lower production cost by microbial rather than mammalian culture.

## Other products of mammalian cell culture

Apart from monoclonal antibody and viral vaccine production, mammalian cell-culture systems are being developed or currently being used to manufacture a

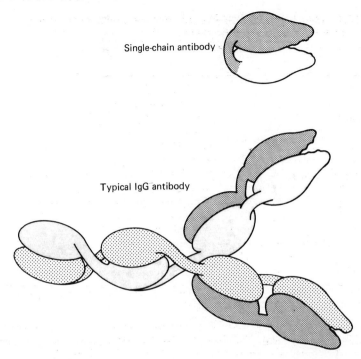

Single-chain antibody

Typical IgG antibody

**Figure 10.16** A new single-chain antibody consisting of the variable portion of a light chain and a variable portion of a heavy chain connected by a peptide. The antigen binding sites are on the right-hand side of each structure (reproduced with permission from Klausner, 1986).

variety of other products especially for human and animal health care and diagnosis. Some of the pharmaceutical products and their applications are listed in Table 10.1. It is predicted that the world market for the human hormones, insulin, growth hormone, fertility hormone, epidermal growth factor and estrogen should reach $1.9 billion by 1991. Markets for tissue plasminogen activator and for interleukin-2, interferon and other lymphokines are expected to develop rapidly and, with many processes based on animal-cell culture and fermentation of recombinant bacterial strains, these technologies will compete with one another. Large-scale cell-culture techniques involve bioreactors similar to those used for monoclonal antibody production. Example processes for production of interferons by cell culture are provided in the next section.

The development of high-level genetic systems allows genes of interest to be inserted into established cell lines in order to achieve production of compounds such as human interferons, human growth hormone and human plasma factor-VIII. A major advantage of cell culture as a host is that proteins can be processed exactly as they occur in the body (with glycosylation, if required) and can be secreted in the correct conformation into the medium. Difficulties in culturing shear-sensitive cells have been overcome through the development of low-shear

**Table 10.1**　Mammalian cell-culture products and application/treatment
(reproduced with permission from Ratafia, 1987)

| Product | Disease/condition |
| --- | --- |
| Lymphokines | Viral infections |
| Erythropoietin | Anemia, hemodialysis |
| Recombinant insulin | Diabetics, insulin-dependent |
| Beta cells | Diabetics |
| Urokinase | Blood clots |
| Granulocyte stimulating factor | Wounds, severe |
| Tissue plasminogen activator | Heart attacks, for survival to hospital |
| Transfer factor | Multiple sclerosis |
| Protein C | Hip surgery, protein C deficiency |
| Epidermal growth factor | Burns |
| Factor VIII | Hemophilia |
| Human growth hormone | Pituitary deficiency |
| Orthoclone | Kidney transplant rejection |
| Alpha-interferon | Hairy-cell leukemia |

bioreactors having good environmental control. Cost disadvantages of using sera-containing media are being overcome through the development of defined media.

**Anti-cancer agents**

Cancer chemotherapy has been directed at the discovery of cytotoxic agents capable of inhibition of mammalian cell division. The total world market for anti-cancer agents was in excess of $1 billion in 1987 and is projected to increase to $27 billion by the end of the century. Fermentation-derived agents account for 30–45% of the current anti-tumour drug market. Many anti-cancer agents produced by micro-organisms and transformed mammalian cells are in the early stage of drug evaluation. In this section examples of the biotechnology of production of cytotoxic drugs by micro-organisms and animal cells will be reviewed.

Adriamycin, one of a group of anthracycline antibiotics produced by *Streptomyces peucetius* has led the sales of all anti-cancer agents in the USA for some years. Enzymes which deplete nutritionally essential and non-essential amino acids have also been produced commercially for treatment of human leukemias and solid tumours. L-Asparaginase has been the most successful anti-neoplastic therapeutic enzyme and asparaginase from guinea-pig serum is preferred to the *E. coli* or *Erwinia carotovora* enzymes because of its reduced antigenicity. The demand for agents such as interferon and monoclonal antibodies for treatment of lung, breast, colon and prostate tumours and certain types of leukemias and lymphomas is predicted to grow at the fastest rate in the coming years. Monoclonal antibodies are currently undergoing clinical evaluations.

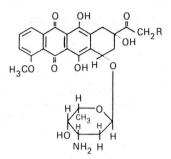

Daunorubicin  R = H
Adriamycin    R = OH

**Fig. 10.17**   Structures of daunorubicin (R = H) and doxorubicin (Adriamycin) (R = OH).

ANTHRACYCLINES

Daunorubicin, the first clinically-effective anthracycline antibiotic was isolated from *Streptomyces coeruleorubidus* in 1963. It has been used primarily for treatment of acute leukemia. Doxorubicin (known commercially as Adriamycin) was obtained in an attempt to improve daunorubicin, by mutation of *S. peucetius*. These anthracyclines exhibit dose-dependent cumulative and irreversible cardiac toxicity and consequently many other anthracyclines were isolated or produced semi-synthetically in an attempt to obtain an effective agent with reduced cardiotoxicity. Their structures are illustrated in Fig. 10.17.

A fermentation process for production of daunorubicin is summarized in Fig. 10.18. Production media contain carbon sources such as starch or glucose, complex nitrogen sources such as soya flour, distiller's solubles, yeast extract and fishmeal and inorganic salts. Aeration is maintained at 0.5 vvm and impeller tip speed is set at 750 ft. min$^{-1}$ in scale-up and production fermenters. Daunorubicin is produced in the form of glycosides. Acidification of the whole broth after fermentation lyses the cells and hydrolyses the glycosides to daunorubicin.

Because of their potential toxicity, exposure to liquids containing cytotoxic material during production must be avoided by use of appropriate containment technology in production and purification.

INTERFERON

Systems for production of interferon from diploid fibroblasts and leukocytes provide examples of recent mammalian cell-culture processes for production of anti-cancer agents.

Diploid fibroblasts are incubated in tissue culture flasks and roller bottles to achieve confluent growth. These cells may be induced to produce human fibroblast interferon (HuIFN-$\beta$) by exposure to poly I:C (polyinosinic-polycytidylic acid) and yields are increased by treatment with 100 U ml$^{-1}$

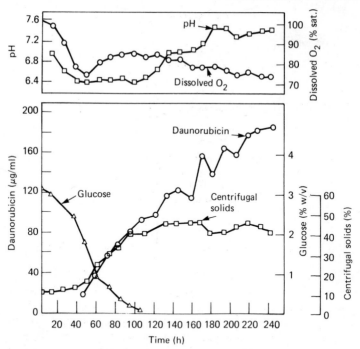

**Fig. 10.18** Production of daunorubicin by submerged culture of *Streptomyces peucetius* (reproduced with permission from Flickinger, 1985).

HuIFN-$\beta$ as a primer prior to induction with poly I:C. The cells are anchorage-dependent and provision of an adequate surface area for growth is a major obstacle to scale-up. Methods used have included use of stationary flasks, roller bottles, stacked plates, fibres and microcarriers in stirred tank reactors. Inocula for successive steps are generally 20–25% and final cell concentrations of $2$–$5 \times 10^6$ cells ml$^{-1}$ may be obtained. Typical yields of HuIFN-$\beta$ are $3.2 \times 10^4$ U ml$^{-1}$.

Leukocyte interferon (HuIFN-$\alpha$) is produced from suspensions of leukocytes using HuIFN-$\alpha$ as primer and Sendai or Newcastle disease virus as inducer. The 'buffy coat' leukocytes are recovered from whole blood and purified free of leukocytes by use of ammonium chloride. Reported yields of HuIFN-$\alpha$ are $20\,000$–$60\,000$ U ml$^{-1}$. An outline production process is presented in Fig. 10.19.

While results with interferon as an antitumour drug in solid tumours have not been particularly encouraging, IFN has proved valuable in therapy of some rarer types of cancer, notably hairy cell leukemia.

### Other microbial pharmacological agents

The research tools of molecular biology are providing opportunities for discovery of new generations of microbial pharmacological products and also for the

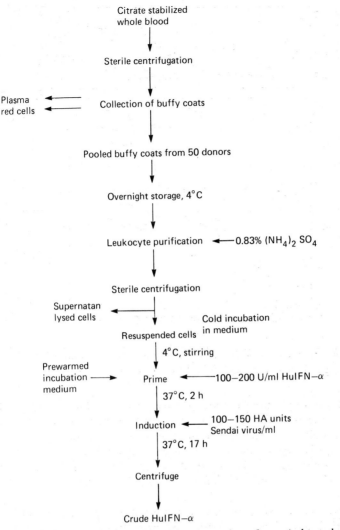

**Fig. 10.19** Process for production of human α-interferon (adapted with permission from Flickinger, 1985).

identification of new bioactivities of previously characterized microbial products. Product categories include anti-inflammatory agents, enzyme inhibitors, immunodepressants, immunostimulants, hormone agonists and antagonists. Structures of some of these compounds are related to structures of some of the antibiotics. Many are naturally produced or chemically modified secondary metabolites and have a variety of modes of action including receptor binding, cell wall or membrane interaction, nucleic acid binding, enzyme activation or inhibition. The use of efficient *in vitro* bioactivity screening programmes is considered a prerequisite to successful discovery of these new agents. Gene cloning

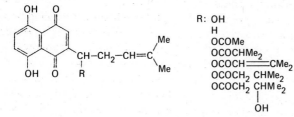

| R: | OH | shikonin |
|----|----|----------|
| | H | deoxyshikonin |
| | OCOMe | acetylshikonin |
| | OCOCHMe$_2$ | isobutylshikonin |
| | OCOCH═══CMe$_2$ | $\beta$ . $\beta$-dimethyl acrylshikonin |
| | OCOCH$_2$ CHMe$_2$ | isovalerylshikonin |
| | OCOCH$_2$ CHMe$_2$ | $\beta$-hydroxyisovalerylshikonin |
| |          | OH | |

**Fig. 10.20**  Structure of shikonin and its derivatives.

facilitates construction of cell based systems designed to detect molecules with new biochemical activities. Receptors and ligands become readily accessible through eukaryotic gene cloning and expression. Detection systems may be designed around methodology such as enzyme-linked immunosorbent assay. Research effort into discovery of pharmacologically active, non-antibiotic, secondary metabolites is intensifying. Many pharmaceutical companies consider this gathering momentum as the start of a second great era akin to the 'Antibiotic Golden Era' of the 1940–1950s.

## Shikonin production by plant-cell culture

Shikonin and its derivatives (Fig. 10.20) occur in roots of the plant *Lithospermum erythrorhizon*. This plant has traditionally been used in Japan as a medicine, due to the antibacterial and anti-inflammatory effects of shikonin. Shikonin, which is a bright red naphthoquinone pigment, is also used as a dye in cosmetics. The plant was imported to Japan from China and Korea for preparation of pure shikonin which has a value of $4500/kg.

**Table 10.2**  Medium and other parameters for production of shikonin derivatives by suspension cultures of *Lithospermum erythrorhizon* (from Fujita and Hara, 1985)

| *Medium (mg/l)* | | | |
|---|---|---|---|
| Ca(NO$_3$)$_2$·4H$_2$O | 1388 | CuSO$_4$·5H$_2$O | 0.6 |
| KNO$_3$ | 160 | Na$_2$MoO$_4$·2H$_2$O | 0.004 |
| KCl | 130 | NaFe-EDTA·3H$_2$O | 3.6 |
| NaH$_2$PO$_4$·2H$_2$O | 38 | KI | 1.5 |
| MgSO$_4$·7H$_2$O | 1500 | Na$_2$SO$_4$ | 2 960 |
| ZnSO$_4$·4H$_2$O | 6 | Sucrose | 40 000 |
| H$_3$BO$_3$ | 9 | 3-Indoleacetic acid | 3.5 |
| *Culture parameters* | | | |
| Incubation temperature (°C) | 25 | Cell yield (g/l) | 17.5 |
| Inoculum concentration | | | |
| (g dry wt/l) | 5.6 | Shikonin cell content (%) | 13.2 |
| $K_La$(h$^{-1}$) | 12 | Shikonin yield (mg/l) | 2 300 |

A cell line was established which could accumulate up to 15% of the biomass as product and productivity was optimized through a cell-culture development programme. The production process consists of a two-stage fermentation. In the first stage for development of biomass, cells are grown in a 200-litre fermenter for about nine days. The cells are harvested and then transferred to a 750-litre shikonin production vessel and incubated for about 14 days. Yield of shikonin derivatives is very sensitive to interrelationships between oxygen supply, amount of inoculum and concentrations of constituents in the medium. Plant-cell secondary metabolites are not secreted into the medium but are rather retained in vacuoles and organelles. Shikonin derivatives may be extracted from the cells using ethanol. Production conditions which produced yields of 2.31 g shikonin derivatives per litre of culture are summarized in Table 10.2.

# Chapter 11

# *Industrial enzymes*

## Applications of enzymes

The bulk of enzymes used in industry are extracellular enzymes of microbial origin. Proteases constitute about 50% of the microbial enzyme market. Use of alkaline serine protease from *Bacillus licheniformis* in detergents is the dominating commercial application of proteases, followed by the use of *Mucor miehei* rennet in cheese manufacture. Applications of *Aspergillus oryzae* fungal protease in the modification of dough for bread- and cracker-making is another significant industrial protease application.

The principal applications of extracellular starch-degrading enzymes are in the conversion of starch to glucose-, maltose- and oligosaccharide-containing syrups, in production of fermentable sugars in brewing and distilling and in the modification of baking flour. Important commercial enzymes used in one or more of these applications are α-amylases from *Bacillus licheniformis*, *Bacillus amyloliquefaciens* and *Aspergillus oryzae* and amyloglucosidase from *Aspergillus niger*.

Pectinases and hemicellulases are produced mainly by strains of *A. niger*. Major components of pectinases are pectin esterase, endopolymethylgalacturonate lyase and polygalacturonase. In fruit-juice extraction and wine-making processes, pectinases increase juice yield, reduce viscosity, improve colour extraction from fruit skins and macerate fruit and vegetable tissue.

Commercial cellulases, produced by strains of *Trichoderma reesei*, *Penicillium funiculosum* and *Aspergillus niger*, currently have limited applications in food processing, brewing, alcohol production and waste treatment. Their major commercial potential is in the conversion of wood lignocellulose and cellulose to glucose. The most important cellulase components are cellobiohydrolase, which has the greatest affinity for crystalline cellulose, endo-β-1,4-glucanase and cellobiase. *Penicillium emersonii* produces a $\beta$-1,3;1,4-glucanase which has important applications in hydrolysis of barley $\beta$-glucans.

Other important microbial extracellular enzymes include yeast and fungal lipases, fungal lactases and dextranases.

Glucose isomerase is the predominant commercial intracellular enzyme used in the food industry, which converts glucose to fructose in the production of high fructose corn syrup. Commercial yeast and bacterial lactases produced by *Kluyveromyces lactis*, *Kluyveromyces fragilis* and *Bacillus* spp. are also of intracellular origin. Intracellular enzymes are also used as clinical diagnostic reagents in genetic engineering R & D and in the biotransformation of chemicals and pharmaceuticals. Important examples of clinical diagnostic microbial enzymes are hexokinase, glucose oxidase, glucose-6-phosphate dehydrogenase and cholesterol oxidase.

Biotransformations have been defined as selective, enzymatic modifications of pure compounds into specific final products or intermediates. They can involve use of isolated enzymes or whole cells used in free or immobilized form. There are many examples of enzymes used in biotransformation reactions. Penicillin acylase is used to hydrolyse penicillin side-chains to produce 6-aminopenicillanic acid (6-APA) and L-amino acid acylases are used in the preparation of resolved L-amino acids from acylated racemic D,L-amino acid mixtures. Only the L-form is deacylated, leaving the D-form which is separated, re-racemized and re-used as substrate. Perhaps the most important groups of enzymic biotransformations are carried out by the pharmaceutical industry, namely the sterol transformations (see Chapter 10).

The most notable non-microbial enzymes are papain, from the plant *Carica papaya*, used in brewing and food processing, and chymosin, extracted from the fourth stomach of unweaned calves, used in cheese-making. In addition to these, malted cereals, used in brewing, provide a source of $\alpha$- and $\beta$-amylases, proteases and $\beta$-glucanases for liquefaction, saccharification and extraction of fermentable sugars and nutrients.

The vast majority of extracellular microbial enzymes are used in the form of free soluble enzymes and are not recovered for re-use. The enzymic treatment cost per batch of substrate is relatively low and enzyme recovery or use of immobilized enzyme systems have in general so far not been cost-effective. In contrast, a considerable number of reactions, carried out in industry using intracellular enzymes, involve use of immobilized enzymes or free or immobilized whole cells, which can be recycled. Commercial glucose isomerases are prepared by immobilization of cells containing the enzyme or by immobilization of the isolated enzyme on synthetic supports, depending on the manufacturer.

Overall costs of production of intracellular enzymes are substantially higher than those for extracellular enzyme production, because of the significantly greater isolation and purification costs and consequently re-use of the enzyme through immobilization is often economically desirable. A major advantage of using free or immobilized cells in biotransformations is that enzyme isolation costs are avoided. However, the cell system must allow adequate rates of penetration and diffusion of substrate into, and product from, the cells and reactions involving formation of undesirable by-products may have to be inhibited or minimized. Cell biotransformation systems may be particularly applicable when the enzyme involved is membrane-associated or requires specific co-factors or where more than one enzyme is involved.

Systems, which involve use of isolated immobilized enzymes or immobilized cells, are advantageous for production of high-purity products free of residual enzyme protein. Enzyme immobilization or retention of the enzyme in immobilized cells generally increases enzyme stability.

**New developments**

DAIRY PROCESSES

Microbial rennets, which are cheaper than calf rennet (chymosin) are used for production of more than one-third of cheese on the world market. The first microbial rennets were relatively heat-stable compared to calf rennet with the result that they tended to remain active in cheese, causing off-flavours and yield reductions. Chemical treatment of *M. miehei* rennet with oxidizing agents such as $H_2O_2$ produced an enzyme having similar temperature properties and producing cheese of similar quality and yield to chymosin. Many biotechnology companies and research institutes have engaged in research to produce chymosin in genetically engineered bacteria and fungi.

Lactose present in milk and whey is less soluble and less sweet than its component monosaccharides, glucose and galactose. Although significant applications for soluble or immobilized lactase exist (Table 11.1) substantial commercial development of these processes has been hampered by the high cost of the enzyme. To overcome this problem Tetra Pak International developed the Tetra Lacta® System, where a low enzyme dose is added to sterilized milk and over 80%

**Table 11.1**  Major applications of lactases

| Raw material | Product | Comments |
|---|---|---|
| Whey | Animal feeds | Allows more whey to be incorporated |
| | | Prevents crystallization of lactose in whey concentrate |
| | Lactose-hydrolyzed whey syrup | Used as an ingredient in bakery, confectionery and ice-cream products |
| Deproteinized whey | Lactose-hydrolyzed permeate syrup | Properties similar to glucose syrups of medium dextrose equivalent |
| Milk | Lactose-hydrolyzed milk | Increases sweetness |
| | | Prevents crystallization in milk concentrates |

lactose hydrolysis takes place in the packed milk during a storage period of one to two weeks.

Industrial enzymes may be added to cheeses to augment the effect of enzymes secreted by starter cultures in cheese-flavour development. Protease from *B. amyloliquefaciens* promotes ripening of cheddar cheese and lipases develop stronger flavours in Italian cheeses. Recently the enzyme sulfhydryl oxidase has been developed to eliminate undesirable thiol flavours which form in UHT preserved milk.

### BREWING AND STARCH HYDROLYSIS

In normal beer, one-third to one-quarter of the wort carbohydrate is present as limit dextrins which are not fermented and remain in the final product. These dextrins are hydrolysed to fermentable sugar by *A. niger* glucoamylase which means that a more dilute wort, having a lower calorie content, may be used to produce the same level of alcohol present in normal beer. The enzyme may be added at a high dose rate to the brew at a wort temperature stand of 65°C prior to wort boiling. Alternatively the enzyme may be added at a low dose rate to the fermenter and allowed to act during the fermentation process. The disadvantage of the latter approach is that normal pasteurization temperatures do not inactivate the enzyme which can cause beer stability problems during storage. One possible solution to this problem is to immobilize the glucoamylase.

A novel recent approach to the use of enzymes in brewing has been to attempt having the fermenting yeast excrete these enzymes during fermentation. *Saccharomyces diastaticus* has been classified as distinct from *S. cerevisiae* in that it produces glucoamylase. However, *S. diastaticus* is unsuitable for brewing since it produces off-flavours. Its glucoamylase enzyme is also too temperature-stable. Another yeast, *Schwanniomyces castelli*, possesses glucoamylase which has the important characteristic of being inactivated at normal pasteurization temperatures. The objective would be to genetically engineer this enzyme into *S. cerevisiae*. Other ongoing research involves cloning β-glucanase, chill-proofing proteases and other brewing enzymes into *S. cerevisiae*.

During beer-maturation processes, oxidative transformation of α-acetolactate into diacetyl and diacetyl reduction to acetoin are rate-limiting for the overall process (Fig. 11.1). The introduction of the enzyme acetolactate decarboxylase to

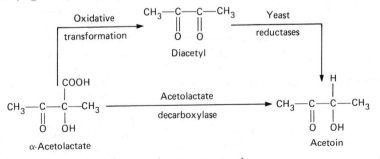

**Fig. 11.1** Conversion of α-acetolactate to acetoin.

freshly fermented beer bypasses the formation of diacetyl and makes it possible to substantially accelerate the maturation process and suitable sources of this enzyme are being sought.

Glucoamylase is used together with high-temperature stable α-amylase in the liquefaction/saccharification of starch to glucose. Although glucoamylase hydrolyses both the α-1,4- and α-1,6-glycosidic linkages, the α-1,6-linkages are only slowly cleaved by this enzyme. Use of debranching enzymes, such as pullulanase, which preferentially hydrolyse the α-1,6-linkages, have been shown to significantly increase saccharification rates. The first pullulanase introduced commercially, from *Aerobacter aerogenes*, had a pH optimum around 6.0 and was not very effective at pH 4.0–4.4 required for saccharification by glucoamylase. Recently, a commercial pullulanase preparation from *Bacillus acidopullulyticus* has been introduced which has the right temperature and pH activity/stability characteristic for saccharification processes.

DEGRADATION OF PLANT CELL-WALL POLYSACCHARIDES

Significant developments are taking place in enzymatic processes for degradation of non-lignified plant tissue such as fruits and vegetables. The cell walls are composed of cellulose fibres and associated hemicelluloses embedded in a matrix of pectic substances. Conventional commercial hemicellulases and cellulases were unable to completely liquefy these complex cell-wall materials. A wide range of enzymes including cellulases, other glucanases, galactanases, arabinases and pectinases are required for full degradation. A greater understanding of the properties of these enzymes and the organisms which produce them will offer the potential for production of a variety of enzyme formulations which would facilitate (a) targeting of enzymes towards degradation of specific tissues while leaving others unaltered, (b) total maceration and solubilization of tissues and (c) total degradation of polysaccharides to monosaccharides.

The enzymatic degradation of lignocellulose is not yet economically feasible, mainly because of the resistance of the lignin component to biological attack. Since crystalline cellulose in wood is embedded in lignin, it is protected from hydrolysis by cellulases. Consequently economical methods must be devised to break up these lignocellulose complexes so that the cellulose is rendered amenable to enzymatic attack. Cellulose hydrolysis requires the synergism of three important enzyme components: cellobiohydrolase to act on microcrystalline cellulose; endo-glucanase to attack amorphous regions in cellulose fibres and more soluble 1,4-β-glucans; and β-glucosidase to hydrolyse cellobiose, the product of the first two enzymes, to glucose. In the most effective commercial enzyme preparation, produced by *Trichoderma reesei*, the activity of the first two enzymes is inhibited by cellobiose. Furthermore, this organism only produces low levels of β-glucosidase which, by causing removal of cellobiose, would reduce its inhibitory effect. The β-glucosidase is in turn inhibited by its product, glucose. These product-inhibition properties will impede the development of enzymatic processes for production of glucose from cellulose. Clearly there is a need to identify or engineer cellulases which are insensitive to product inhibition.

**Table 11.2** Bioconversion of organic compounds in water-immiscible organic solvents

| Substrate | Biotransformation | Catalyst | Organic solvent |
|---|---|---|---|
| Cholesterol | Oxidation of hydroxyl group, isomerization | *Nocardia* cells | Carbon tetrachloride |
| 3β- or 17β-Hydroxysteroids | Oxidation of 3β- or 17β-hydroxyl groups | β-Hydroxysteroid dehydrogenase | Ethyl or butyl acetate |
| 20-Ketosteroids | Reduction of 20-keto groups | 20 β-Hydroxysteroid dehydrogenase | Ethyl or butyl acetate |
| Oestrogens | Oestrogen oligomer formation | Fungal laccase | Ethyl acetate |
| 4-Androstendione | Δ¹-Dehydrogenation | *Nocardia rhodocrous* cells | Benzene–heptane |
| 1,7-Octadiene | 7,8-Epoxidation | *Pseudomonas oleovorans* cells | Cyclohexane |
| N-Acetyl-AA + Ethanol (AA = tryptophan, tyrosine, phenylalanine) | Esterification | Chymotrypsin | Chloroform |
| Glycerol + phosphate | Esterification | Alkaline phosphate | Chloroform |

**Table 11.3**  Advantages of two-phase water-organic solvent systems in enzyme bioconversions

---

- Efficient conversions of substrates poorly soluble in water may be achieved
- Reactions catalyzed by hydrolytic enzymes can be adjusted to favour synthesis
- Products can be easily separated from enzymes, if they are poorly water-soluble
- Enzymes are relatively stable under these conditions and may be further stabilized by immobilization
- Microbial contamination is suppressed by the organic phase

---

ENZYME REACTIONS INVOLVING ORGANIC PHASES

In recent years exciting applications of enzyme reactions, carried out in two-phase water–organic solvent systems and in non-aqueous organic media, have been investigated. Enzymes are usually employed in aqueous solutions and where water itself is a reagent in the enzyme reaction the equilibrium position of the reaction generally favours hydrolysis. By use of organic phases, water can be made limiting and the reaction under these conditions can be shifted towards synthesis.

Two-phase systems, consisting of water and an organic poorly water-miscible solvent, can improve the performance of enzyme-catalysed reactions involving substrates which are relatively insoluble in water. The water phase contains the enzyme and any water-soluble co-factors or co-substrates and the organic phase contains the substrate. The enzyme reaction takes place at the boundary between the phases and this interfacial area may be increased by agitation to form an emulsion. Some examples of enzymic reactions which have been carried out in two-phase systems are provided in Table 11.2. Advantages of using these systems are listed in Table 11.3.

While conventional wisdom holds that enzymes work only in aqueous solutions, recent studies on the ability of enzymes to act as catalysts in organic solvents suggest that most, if not all, enzymes can work in organic solvents. All of the non-covalent interactions which maintain the catalytically active structure of an enzyme, including van der Waals forces, hydrogen bonds and salt-bridges, require water either directly or indirectly and if water is totally removed the enzyme is denatured. However, because water is in nature bound very tightly to the enzyme a layer of essential water exists as a shell around the enzyme. As long as enzymes are surrounded by this monolayer of water, they can function in an organic medium. Unlike the two-phase systems described above, these systems are monophasic, operating in an essentially anhydrous organic medium (that is with a water content of less than 1%). The nature of the solvent is crucial for maintaining the shell of essential water around the enzyme. Best solvents are the most hydrophobic ones, such as hydrocarbons. Solvents that are less hydrophobic will have a higher affinity for water and may remove essential water from the enzyme. Since enzymes are insoluble in almost all organic solvents they form suspensions and operate as stirred suspensions in these monophasic organic systems. Some enzymes exhibit enhanced thermal stability in organic solvents. While pancreatic lipase is almost instantaneously inactivated at 100°C in water, its half-life at 100°C in organic media containing 1% and 0.02% water was 10 min and 10 h, respectively.

When placed in organic media, enzymes have been shown capable of catalysing many reactions which, because of reaction equilibria, are virtually impossible in water. For example, lipases catalyse transesterification, esterification, aminolysis, acyl exchange, thiotransesterification and oximolysis reactions in organic media whereas hydrolysis predominates under aqueous conditions.

These new developments in enzyme technology offer enormous scope for expanding the industrial applications of enzymes especially in the area of organic chemical synthesis.

## Enzyme production

### GENERAL PRODUCTION PROCESSES

Microbial enzymes are produced from high-yielding strains by fermentation under controlled conditions in surface or submerged culture. The various processing stages are illustrated in Fig. 11.2. Extracellular enzymes are secreted into the medium by the cells and the first stage of enzyme recovery from submerged cultures involves separation of the cell-free liquor, containing the enzyme, by filtration or centrifugation. Surface-culture fermentation processes are widely used in addition to submerged culture for production of fungal extracellular enzymes and in this case the post-fermentation semi-solid mass is usually extracted with water prior to filtration. The supernatant or filtrate containing the recovered enzyme(s) is concentrated and sold as a standardized liquid enzyme preparation containing preservatives and/or stabilizers, or precipitated (sometimes prilled), dried and ground for preparation of powdered and granulated enzymes.

In general, recovery of industrial extracellular enzymes does not involve an enzyme fractionation step. Indeed many extracellular enzyme culture broths contain a variety of contaminating enzymes, in addition to the major activity, which improve the application performance of the product. For example, proteases, hemicellulases, cellulases and other enzymes, present in fungal amyloglucosidase preparations, can increase alcohol yields obtained in α-amylase/amyloglucosidase-treated cereal fermentation mashes. In a small number of cases, contaminating enzymes reduce the effectiveness of extracellular enzymes in specific industrial applications and must be denatured or removed, usually from the culture filtrate. Contaminating transglucosidase, present in crude amyloglucosidase preparations, catalyses the conversion of glucose to isomaltose and panose during processes involving starch saccharification. This contaminant obviously reduces the yield of glucose produced and in alcoholic fermentations reduces final ethanol yield, since isomaltose and panose are not fermented to alcohol. Transglucosidase may be removed from amyloglucosidase preparations by adsorption using bentonite. Amyloglucosidase for use in preparation of low carbohydrate beer should not contain appreciable protease activity,

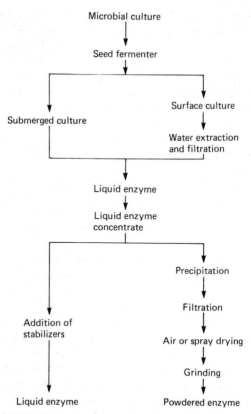

**Fig. 11.2** Flow diagram for production of industrial enzymes by fermentation.

which can have the effect of reducing the head of the final beer product and may produce excess α-amino nitrogen which would encourage contamination during beer storage.

Fermentation processes for production of intracellular enzymes must be terminated prior to the onset of cell lysis which would result in enzyme loss to the medium. Isolation of intracellular enzymes involves cell breakage followed by use of an appropriate combination of biochemical purification techniques to recover the intracellular enzyme to the desired specification. Clinical enzymes are often used to determine substrate concentrations using coupled assay procedures involving a number of enzymes and co-factors, with ultimate measurement of oxidized or reduced NAD or NADP at 340 nm. While specifications for these enzymes include minimum acceptable enzyme-specific activity values, much more important elements in the specification relate to maximum allowable levels of contaminating enzymes which would give false assay results. For example, glucose may be measured using a hexokinase/glucose-6-phosphate dehydrogenase-coupled assay system. Oxidation of glucose results in the quantitative production of reduced NADH which may be measured spectrophotometrically. In this case, contaminating glucose oxidase, glucose-6-phosphate isomerase, 6-phosphogluconate dehydrogenase, ATP-ase and NADH oxidase would interfere with the coupled enzyme assay and therefore should be present in negligible amounts in the diagnostic enzyme preparations being used.

Processes for production of micro-organisms for cellular biotransformations must take a number of considerations into account. Enzyme activities causing undesirable side-reactions or degradation of the desired product should be minimized. In cases where the undesirable enzymes are not required for cellular growth and metabolism, mutation selection techniques may be used to eliminate them from the organism. Where these undesirable enzymes are required for growth, physical (heat) or chemical (acid, alkali, solvents, inhibitors, detergents) treatment methods are used to suppress enzyme activity by inactivation or inhibition prior to the biotransformation step. Finally, it may be necessary to modify cell permeability to allow adequate rates of diffusion of substrate and product to and from the site of enzyme activity. Sometimes this is achieved by disruption of the membrane with acetone on toluene.

ENZYME IMMOBILIZATION PROCESSES

Enzyme immobilization within the host cell was the technique used for the first commercial process involving glucose isomerase. The enzyme-containing *Streptomyces* cells were heated for short periods to denature autolytic enzymes, to stabilize the cells and to make them permeable to small molecules. Another industrial method involved the use of flocculating agents to fix glucose isomerase within *Arthrobacter* cells. These cell preparations were then pelleted for use in enzyme reactors. Glucose isomerase may also be fixed within cells using a bifunctional aldehyde cross-linking reagent such as glutaraldehyde. In another process for production of industrial immobilized glucose isomerase the enzyme-containing microbial cells are fixed with glutaraldehyde and entrapped in gelatin.

Fibre-entrapment processes, involving materials such as cellulose triacetate, have also been used to trap isolated enzymes for industrial use. Enzymes immobilized for commercial application in this way include glucose isomerase, aminoacylase and β-galactosidase. Immobilized glucose isomerase has also been prepared commercially by adsorption of an enzyme from *Streptomyces* sp. on DEAE-cellulose and on alumina. The first large-scale application of immobilized enzymes involved use of DEAE-Sephadex bound fungal amino acylase for resolution of chemically-synthesized acyl-D,L-amino acid.

MICROBIAL ENZYME BIOSYNTHESIS

The most important organisms involved in industrial extracellular enzyme production are *Bacillus* spp. and *Aspergillus* spp. These two genera account for 80–85% of the extracellular enzyme market. *Trichoderma reesei* is currently the best industrial producer of cellulases. While the existing market for cellulases is small, the commercial potential for the future may be extremely large, perhaps outstripping all other enzymes combined, if current technical problems related to lignocellulose hydrolysis can be overcome. Glucose isomerases are produced for fructose syrup manufacture by *Arthrobacter*, *Streptomyces* and *Actinoplanes* species.

General aspects of induction and repression of enzyme synthesis and mechanisms of enzyme secretion have been discussed in Chapter 2. A feature of many extracellular enzymes is the extended stability of their mRNAs. This may be associated with the need to transport the mRNA from its site of synthesis to translation sites on the inner surface of the cytoplasmic membrane. The observation that cells producing proteases appear to be capable of accumulating large pools of protease-specific mRNA has been in part attributed to the relatively long half-life of protease-specific mRNAs.

Commercially important amylases and proteases of *B. amyloliquefaciens* and *B. licheniformis* are produced constitutively, that is, the rate of enzyme synthesis is relatively constant, irrespective of the substrate in the fermentation medium. Inducible extracellular enzymes tend to be synthesized at a low basal rate when the substrate is absent, but the rate of synthesis may increase several thousand-fold on induction. While maltose appeared to be the best inducer of the α-amylase of *A. oryzae*, starch, glucose and a variety of other α-D-glucosides also induce the enzyme. Starch, maltose and glucose all stimulate production of amyloglucosidase by *A. niger*. Clearly these enzymes are not catabolite-repressed by glucose. While cellulose induces the cellulase complex of *T. reesei*, the true inducer is probably cellobiose. Lactose is an inducer of *T. reesei* cellulases and also serves as a carbon source for growth. *Trichoderma* spp. and *Aspergillus* spp. produce extracellular cellobiases (β-glucosidases) which are induced by many β-glucosides including cellulose but not cellobiose. These organisms also constitutively synthesize intracellular β-glucosidases. Polygalacturonase production by *A. niger* is enhanced by the presence of pectic substances in the growth medium.

Synthesis of many extracellular enzymes is controlled by end-product or catabolite repression. The α-amylases of some strains of *B. licheniformis* are catabolite-repressed by major end-products of α-amylase starch hydrolysis. Glucose causes catabolite repression of *T. reesei* cellulases. Glucose represses the synthesis of cellobiohydrolase and endoglucanase in the presence of cellulose and other inducers. The constitutive intracellular β-glucosidases produced by *Trichoderma* and *Aspergillus* species are relatively insensitive to catabolite repression. Production of polygalacturonase by *A. niger* is catabolite-repressed by glucose. While protease synthesis in *Bacillus* species is constitutive, it is also sensitive to end-product repression. Glutamate and aspartate repress protease synthesis by *B. subtilis* and protease secretion by *B. licheniformis*.

The classical operon regulation models for enzyme induction and repression relate to *Escherichia coli* (see Chapter 2). Nevertheless, present evidence suggests that control of extracellular enzyme synthesis is exerted primarily at the level of transcription and that the operon model for gene regulation is applicable to these systems and indeed has been confirmed in the case of penicillinase synthesis by *B. licheniformis*. In most cases the operons of extracellular enzymes appear to consist of a single structural gene rather than the typical cluster of structural genes coding for a series of enzymes in a metabolic pathway. Whereas cAMP is the major mediator of catabolite repression in enteric bacteria, such molecules are absent from *Bacillus* strains examined to date and the mechanism of catabolite repression remains obscure. It has been speculated that guanosine-3′,5′-cyclic monophosphate may be the effector in *B. licheniformis*. While regulation of transcription appears also to be the principal control point in eukaryotic protein synthesis, gene expression is more complex than in bacteria. A high-molecular-weight RNA transcript is processed, whereby segments (introns) are removed and the remaining sequences (exons) are spliced together to produce the translatable mRNA which is further modified at each end.

Extracellular enzyme synthesis is usually associated with the exponential or post-exponential phase of growth. Under most growth conditions, *Bacillus* species produce extracellular proteases during the post-exponential phase of growth. α-Amylase synthesis in *B. licheniformis* and *A. oryzae* occurs during exponential growth whereas in *B. subtilis* and *B. amyloliquefaciens* the enzyme is produced in the post-exponential phase. Where enzymes are produced in the post-exponential phase it has been postulated that de-repression of these enzymes results from modification of RNA polymerase specificity, through modulation of a small effector molecule or through compositional changes in the RNA polymerase, sometimes associated with events such as sporulation.

## Recombinant enzyme products

The techniques of genetic engineering have resulted in the development of a number of novel products in the pharmaceutical industry. Since enzymes are direct gene products they are good candidates for improved production through genetic engineering. The benefits of achieving commercially useful processes are expected to include: (a) cost savings in enzyme production; (b) production of enzymes in GRAS ('generally regarded as safe') organisms suitable for use in food processes and (c) specific genetic modifications at the DNA level to improve enzyme properties such as thermal stability and other performance characteristics.

A high level of expression with many foreign proteins/enzymes has been achieved in *E. coli*. Many foreign proteins accumulate in *E. coli* cytoplasm in the form of granules. While the granule is simply isolated to purify the cloned recombinant product, dissolution of the granule by detergents or denaturing agents is required with the disadvantage that a refolding step to renature the active native protein may substantially reduce yield while adding to cost. Recent research has placed emphasis on cloning secretion vectors into *E. coli* which enable the foreign protein to be secreted with removal of a signal peptide by the signal peptidases which are present in *E. coli*.

For bulk production of enzymes perhaps the ultimate solution is to genetically engineer useful enzymes from new sources into the current work-horses of the industrial·enzyme industry, *B. subtilis*, *B. licheniformis*, *A. niger*, etc.

Members of the genus *Bacillus* have been successfully used as hosts for production of important proteins produced through recombinant DNA techniques. Difficulties have been encountered with stability of plasmids, as plasmid DNA may not segregate equally between daughter cells during cell division. This can result in formation of plasmidless daughter cells which have a growth advantage since energy need not be expended in product formation. In addition plasmid structural deletions often occur. *Bacillus* cloning hosts should contain reduced protease levels in order to reduce proteolytic degradation of foreign protein products. Once these problems are overcome, *Bacillus* strains may become the ideal hosts for extracellular production of heterologous proteins where post-translational glycosylation is not required.

Filamentous fungi, such as *A. niger*, can produce up to 20 g/l enzyme from a single glucoamylase gene copy. They can be engineered to express and secrete proteins of bacterial origin. It has also been reported that the basic processes of secretion in fungal and mammalian systems are very similar. While a greater understanding of the glycosylation pathways in these eukaryotes is needed, fungal systems offer much potential for synthesis and secretion of large quantities of mammalian proteins.

*Saccharomyces cerevisiae* is capable of expressing, glycosylating and secreting enzymes from fungi and other eukaryotes, for example glucoamylase of *Aspergillus*. However, *Saccharomyces* cannot correctly process the introns in the fungal gene so that an intron-free gene must be used. *Aspergillus* species are, however, much better than yeasts at high-level expression.

Bacteria can express at high levels. Gram-negative bacteria are in general no good extracellular protein secretors and Gram-positive bacteria are significantl; better. Some of the secretion problems may be overcome by proper design of signal sequences. So far the filamentous fungi have not manifested the ability to produce heterologous proteins at the high synthesis rates observed for homologous products.

The successful surmounting of barriers preventing manipulation of recombinant organisms to produce and secrete extremely high yields of heterologous proteins is an essential priority in order that recombinant DNA technology may be applied to production of commodity protein products, including industrial enzymes.

# Chapter 12

# *Waste treatment*

## Introduction

The largest scale at least partially controlled microbial fermentation process involves effluent treatment. Traditionally, the objectives of waste treatment processes have been to reduce the concentration of organic matter in waste water and to reduce the number of potential pathogens in the waste, so that the effluent may be safely discharged into the environment.

The organic content of domestic, agricultural and industrial wastes causes the largest polluting problem as it is metabolized by organisms present in the receiving waters, thereby causing a rapid depletion of oxygen and elimination of the natural aquatic flora and fauna. This organic content of waste may be determined by three tests. The chemical oxygen demand (COD) and total organic carbon (TOC) tests measure total organic carbon by chemical oxidation and pyrolysis, respectively, both using infra-red $CO_2$ detection. The biological oxygen demand ($BOD_5$) test determines the fraction of the waste which is biologically oxidizable by a 5-day incubation using an adapted inoculum. Waste suspended solids measurements, using filtration and turbidity methods, are designed to assess the effects of pollution on light penetration (affecting algal growth) and on solids deposition (affecting river-bed life). While nitrogen is an essential nutrient in waste treatment processes, residual ammonia may be toxic to aquatic life. Oxidized products of ammonia can result in excessive algal growth and in cases where the waste water is recycled for water supply, nitrate can be toxic to growing children.

**Table 12.1**   Stages which may be included in aerobic microbiological waste treatment

1. Adsorption of substrate to biological surface
2. Degradation of adsorbed substrate by extracellular enzymes
3. Cellular absorption of dissolved materials
4. Excretion of breakdown products
5. Cell lysis and ingestion by a secondary cell population

## Waste treatment systems

Waste streams are normally complex in composition and waste treatment plants need to contain a diverse range of micro-organisms having the capability to metabolize various waste materials so that the effluent may be discharged into the environment without adverse effects. Most domestic sewage treatment plants and indeed many industrial chemical and food waste water treatment systems include three or four basic stages: a primary process stage to remove grit and direct other solids to a sludge digester; a secondary aeration stage to degrade dissolved organic compounds and some suspended solids by microbial metabolism; an optional chemical precipitation and phosphorus/nitrogen separation stage; an anaerobic digestion stage to reduce the product sludge volume and to generate methane.

Relatively basic systems, such as oxidation ponds and barrier ditches, exist for disposal of agricultural wastes using aerobic methods. However, particularly for large intensive livestock farms, there has been considerable interest in the development of anaerobic digestion units for digestion of manure wastes.

AEROBIC WASTE TREATMENT

Stages which may be included in aerobic microbiological waste treatment are summarized in Table 12.1. Two main types of systems exist, percolating or trickling filters, and activated sludge processes.

*Percolating filters*
A typical simple percolating filter consists of a cylindrical tank about 10 m in diameter and 2.5 m in depth. The tank, which is packed with a porous bed of stones or other support material, has a bottom draining system (Fig. 12.1). As effluent trickles through the bed, a slime biological layer forms on the surface of the support material. Air flows upward through the bed permitting biological oxidation of the waste materials. Excessive growth of micro-organisms in the filter restricts ventilation and flow, causing filter blockage and failure. Alternating double penetration systems, where the effluent flow is periodically reversed, have improved filter efficiency in terms of BOD removal.

*Activated sludge*
The most common alternative to the trickling filter is the activated sludge process, which involves mixing with aeration and stirring by bottom diffusion or surface

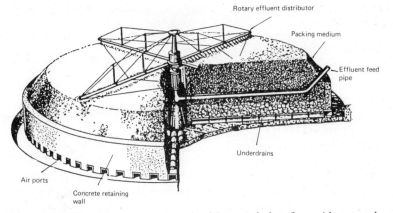

**Fig. 12.1** Trickling filter (reproduced with permission from Abson and Todhunter, 1967).

agitation (see Fig. 3.9 on p. 51). The partially clarified effluent is directed into an eration vessel containing a flocculated suspension of micro-organisms and is .erated during a residence time of 4–10 hours. The effluent then passes to a sedimentation tank to remove the flocculated active sludge, some of which is recycled to the aeration tank to be mixed with the incoming waste. The remainder is dewatered and dried for disposal or use as a fertilizer. Activated sludge is a more powerful waste treatment process than filtration as it can treat much higher effluent loads per unit volume than percolating filters. However, running costs are significantly higher due to the aeration and mixing requirements.

*Recent developments*
The basis of aerobic waste treatment plant design is to optimize the fermentation environment involving the waste, the microbial population and air. Normally aeration efficiency is the limiting factor. Use of light plastic support media, moulded to optimize surface area and ventilation, has facilitated the development of tall space-saving filter plants which do not require substantial retaining walls.

Another recent development is the use of rotating discs which are about 50% submerged so that the microbial film slowly alternates between exposure to liquid and air environments. A typical unit 7.5 m in length and 3.5 m in diameter will have a surface area of 9500 m$^2$ for biological growth. This 'rotating biological contactor' has characteristics similar to percolating filters.

Availability of cheap oxygen has led to the design of a modified enclosed activated sludge plant which uses pure oxygen to achieve higher dissolved oxygen concentrations and operates at higher biomass concentrations and lower residence times without loss of efficiency (Fig. 12.2). An efficient aerated waste treatment plant has been developed using the design principles of the air-lift fermenter containing an internal draft tube (see Fig. 3.10 on p. 52). Incoming waste, returned sludge and process air are injected into the down-flow section. Air bubbles are injected into the up-flow section, thus reducing the density of the

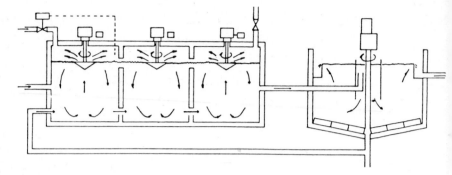

**Fig. 12.2** Enclosed pure oxygen activated sludge plant (reproduced with permission from Wheatley, 1984).

dispersion in that zone. The density gradient between the down-flow and up-flow liquids, above the point of air injection, is responsible for inducing liquid circulation.

Recently fluidized-bed aerobic waste treatment systems have been designed which in essence combine the operating principles of the percolating filter and activated sludge systems. Biological support materials such as plastic or sand are used and fluidization is achieved by injection of air or oxygen at the bottom of the bed (Fig. 12.3). High biomass concentrations can be maintained in the system. These new systems exhibit great promise for the high-intensity removal of wastes.

## ANAEROBIC WASTE TREATMENT

Anaerobic digestion of wastes, which originated with use of septic tanks, now

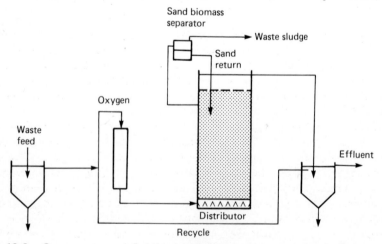

**Fig. 12.3** Oxygen aerated fluidized bed waste treatment plant using sand support particles (reproduced with permission from Wheatley, 1984).

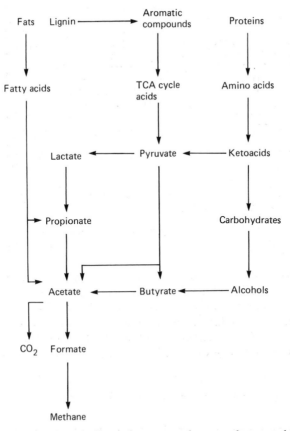

**Fig. 12.4** Anaerobic metabolism of compounds to methane and carbon dioxide.

involves a range of high rate digesters. Some advantages which anaerobic systems have over aerobic systems include the lower power requirement per unit BOD treated, digester load capacity, production of combustible gas as an end-product and production of reduced biomass volume for disposal. Despite these advantages, it is not possible to obtain total degradation of waste using this method and polishing by aerobic methods may be required.

The biochemical sequences involved during anaerobic degradation of waste to methane and carbon dioxide are illustrated in Fig. 12.4. These processes depend on the development of a complex interdependent bacterial community involving three microbial groups: the hydrolytic fermentative group; the hydrogenic acetogenic group; and the hydrogenotrophic methanogenic group. The hydrolytic fermentative bacteria convert complex polymers to sugars, organic acids, alcohols and esters generating carbon dioxide and hydrogen. The hydrogenic, acetogenic bacteria transform the fermentation products of the first group into acetate, carbon dioxide and hydrogen. The methanogens then convert the acetate and hydrogen into methane and carbon dioxide.

The slow growing methanogens, which are susceptible to stress from hydrogen accumulation and from the influent load, cause most of the operational problems associated with anaerobic digestion systems. It is only at hydrogen partial pressures of less than $10^{-3}$ atm, that carbohydrates can be degraded successfully under anaerobic conditions. The hydrogenotrophic methanogens which are responsible for scavenging hydrogen have an estimated doubling time of 8–10 h whereas the acid-forming hydrolytic bacteria have a doubling time of about 30 min. A stressful situation leading to a decline in the methanogen population and a relative increase in hydrolytic bacteria will result in an accumulation of hydrogen. In the presence of an increased hydrogen concentration, the metabolism of the acid-forming bacteria is altered and propionic, butyric, caproic, valeric and lactic acids are produced instead of acetate and hydrogen. Consequently, since digestion overloading is indicated by a rise in concentration of hydrogen and certain acids, continuous monitoring of these metabolites may be used to control the feed rate.

*Anaerobic reactor design*
Some basic anaerobic reactor designs are illustrated in Fig. 12.5. The conventional digester (a) which is a completely mixed system without biomass recycle has a low waste treatment efficiency. Systems (b) to (e) are designed to retain high biomass concentrations within the reactor, (b) by recovery and recycling from the effluent stream, (c) by flocculation, and (d) and (e) by microbial film attachment to support material.

## Microbial inoculants and enzymes for waste treatment

During the past 10 years a series of microbial inoculants and enzyme additives have been developed for waste treatment applications. Commercial microbial enzymes and the micro-organisms that produce them may be added to effluent

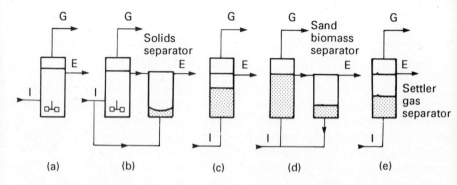

**Fig. 12.5**   Reactor design for anaerobic digestion. (a) Standard digester; (b) anaerobic contact process; (c) anaerobic filter; (d) fluidized bed; (e) upflow anaerobic sludge type. I, Influent; G, gas; E, effluent.

treatment systems to accelerate breakdown of cellulose, hemicellulose, fat, proteins, etc. Microbial inoculants may also be used during start-up of a waste treatment system or to accelerate recovery of a fermentation which has been stressed by an influent load or by a toxic feed. In addition, new genetic techniques are enabling inoculant producers to develop more efficient strains which may aerobically or anaerobically digest waste materials at faster rates than their counterparts present in the natural waste treatment microbial population.

Apart from applications of inoculants in normal municipal and agricultural waste treatment, there is substantial interest in the use of biological systems for degradation of hydrocarbons and recalcitrant chemical compounds. Microbial inoculants are available for degradation of crude oils and selected toxic materials such as phenol. Microbial enrichment culture isolation methods and genetic manipulation investigations are currently being directed toward the development of suitable microbial strains for degradation of compounds including azo dyes, stilbenes, choloraromatics, chlorinated hydrocarbons and DDT.

# Bibliography

Abson, J.W. and Todhunter, K.H. (1967). Effluent disposal, in *Biochemical and Biological Engineering Science*, Vol. 1. Ed. Blakebrough, N., pp. 310–343. London, Academic Press.

Adler-Nissen, J. (1987). Newer uses of microbial enzymes in food processing, *Tibtech* **5**, 170–174.

Aharonowitz, Y. and Cohen, G. (1981). The microbial production of pharmaceuticals, *Scientific American* **245**, 141–152.

Aida, K., Chibata, I., Nakayama, K., Takinami, K. and Yamada, H. (eds.) (1986). *Biotechnology of Amino Acid Production. Progress in Industrial Microbiology*, Vol. 24. Amsterdam, Elsevier.

Anderson, C. and Solomons, G.L. (1984). Primary metabolism and biomass production from *Fusarium*, in *The Applied Mycology of Fusarium*, Eds. Moss, M.O. and Smith, J.E., (British Mycological Society Symposium No. 7) pp. 231-250. Cambridge, Cambridge University Press.

Anderson, L.A., Phillipson, J.D. and Roberts, M.F. (1985). Biosynthesis of secondary products by cell cultures of higher plants, in *Advances in Biochemical Engineering/Biotechnology*, Vol. 31. Ed. Fiechter, A., pp. 1–36. Berlin, Springer-Verlag.

Atkinson, B. and Mavituna, F. (1983). *Biochemical Engineering and Biotechnology Handbook*. London, Macmillan.

Axelsson, H.A.C. (1985). Centrifugation, in *Comprehensive Biotechnology*, Vol. 2. Ed. Moo-Young, M., pp. 325–350. Oxford, Pergamon.

Bacus, J. (1984). Update: meat fermentation 1984, *Food Technology* **38** (6), 59–63.

Bailey, J.E. and Ollis, D.F. (1986). *Biochemical Engineering Fundamentals*, 2nd Edition. New York, McGraw-Hill.

Belter, P.A. (1985). Ion exchange recovery of antibiotics, in *Comprehensive Biotechnology*, Vol. 2. Ed. Moo-Young, M., pp. 473–487. Oxford, Pergamon.

Benda, I. (1982). Wine and brandy, in *Prescott and Dunn's Industrial Microbiology*, 4th Edition. Ed. Reed, G., pp. 293–402. Westport, AVI Publishing Co.

Bernstein, S., Tzeng, C.H. and Sisson, D. (1977). The commercial fermentation of cheese whey for the production of protein and/or alcohol, in *Single Cell Protein from Renewable and Nonrenewable Resources, Biotechnology and Bioengineering Symposium*, **7**. Eds. Humphrey, A.E. and Gaden (Jr.), E.L., pp. 1–9. New York, Wiley.

Berry, D.R. (1984). The physiology and microbiology of Scotch whisky production, in *Progress in Industrial Microbiology*, Vol. 19. Ed. Bushell, M.E., pp. 199–244. Amsterdam, Elsevier.

Best, D.J., Jones, J. and Stafford, D. (1985). The environment and biotechnology, in *Biotechnology, Principles and Applications*. Eds. Higgins, I.J., Best, D.J. and Jones, J., pp. 213–256. Oxford, Blackwell Scientific.

Beuchat, L.R. (1984). Fermented soybean foods, *Food Technology* **38** (6), pp. 64–70.

Boeriu, C.G., Dordick, J.S. and Klibanov, A.M. (1986). Enzymatic reactions in liquid and solid paraffins: application for enzyme-based temperature sensors, *Bio/Technology* **4**, 99–103.

Bonnerjea, J., Oh, S., Hoare, M. and Dunnill, P. (1986). Protein purification: the right step at the right time, *Bio/Technology* **4**, 954–958.

Brierley, C.L., Kelly, D.P., Seal, K.J. and Best, D.J. (1985). Materials and biotechnology, in *Biotechnology, Principles and Applications*. Eds. Higgins, I.J., Best, D.J. and Jones, J., pp. 163–212. Oxford, Blackwell Scientific.

British Valve Manufacturers Association (1966). *Technical Reference Book on Valves for the Control of Fluids*. Oxford, Pergamon Press.

Brock, T.D. (1986). *Thermophiles: General, Molecular and Applied Microbiology*. New York, Wiley.

Buckland, B. (1985). Fermentation exhaust gas analysis using mass spectrometry, *Bio/Technology* **3**, 982–988.

Bull, D.N. (1985). Instrumentation for fermentation process control, in *Comprehensive Biotechnology*, Vol. 2. Ed. Moo-Young, M., pp. 149–163. Oxford, Pergamon.

Bull, D.N., Thoma, R.W. and Stinnett, T.E. (1983). Bioreactors for submerged culture, in *Advances in Biotechnological Processes*, Vol. 1. Eds. Mizrahi, A. and Van Wezel, A.L., pp. 1–30. New York, Alan R. Liss.

Bulloch, W. (1979). *The History of Bacteriology*. New York, Dover.

Bu'Lock, J.D., Detroy, R.W., Hostalek, Z. and Munin-Al-Shakarchi, A. (1974). Regulation of secondary biosynthesis in *Gibberella fujikuroi*, *Transactions of the British Mycological Society* **62**, 377–389.

Bylinsky, G. (1987). Coming: star wars medicine, *Fortune* **115**(9), pp. 153–165.

Cantell, K., Hirvomen, S., Kauppinen, H.L. and Myllyla, G. (1981). Production of interferon in human leukocytes from normal donors with the use of Sendai virus. *Methods in Enzymology*, Vol. 78A. Ed. Pestka, S. pp. 27–38. New York, Academic Press.

Carleysmith, S.W. and Fox, R.I. (1984). Fermenter instrumentation and control, in *Advances in Biotechnological Processes*, Vol. 3. Eds. Mizrahi, A. and Van Wezel, A.L., pp. 1–51. New York, Alan R. Liss.

*Commercial Biotechnology, An International Analysis*, (1984). Office of Technology Assessment Report. Washington, US Congress.

Cooney, C.L. (1981). Growth of microorganisms, in *Biotechnology*, Vol. 1. Eds. Rehm, H.J. and Reed, G., pp. 73–112. Weinheim, Verlag-Chemie.

Cooney, C.L. (1983). Bioreactors: design and operation, *Science* **219**, 728–734.

Crueger, W. and Crueger, A. (1984). *Biotechnology: a Textbook of Industrial Microbiology*. Madison, Science Tech, Inc.

Cullen, D. and Leong, S., (1986). Recent advances in the molecular genetics of industrial fungi, *Tibtech* **4**, 285–288.

D'Amore, T. and Stewart, G.G. (1987). Ethanol tolerance of yeast, *Enzyme and Microbial Technology* **9**, 322–330.

Darnell, J., Lodish, H. and Baltimore, D. (1986). *Molecular Cell Biology*. New York, Scientific American Books.

Deacon, J.W. (1984). *Introduction to Modern Mycology*. Oxford, Blackwell Scientific Publications.

Demain, A.L. (1971). Overproduction of microbial metabolites and enzymes due to alteration of regulation, in *Advances in Biochemical Engineering*, Vol. 1. Eds. Ghose, T.K. and Fiechter, A., pp. 113–142. New York, Springer-Verlag.

Demain, A.L. (1980). The new biology: opportunities for the fermentation industry, in *Annual Reports on Fermentation Processes*, Vol. 4. Ed. Tsao, G.T., pp. 93–208. Orlando, Academic Press.

Demain, A.L. and Solomon, N.A. (1981). Industrial microbiology, *Scientific American* **245**, 67–75.

Demain, A.L. and Solomon, N. (1985). *Biology of Industrial Microorganisms*. California, Benjamin/Cummings Co.

Dimmling, W. (1985). Critical assessment of feedstocks for biotechnology, in *Critical Reviews of Biotechnology*, Vol. 2. Eds. Stewart, G.G. and Russell, I., pp. 233–285. Boca Raton, CRC.

Doelle, H.W., Ewings, K.N. and Hollywood, N.W. (1982). Regulation of glucose metabolism in bacterial systems, in *Advances in Biochemical Engineering*, Vol. 23. Ed. Fiechter, A. pp. 1–35. Berlin, Springer-Verlag.

Dwyer, J.L. (1984). Scaling up byproduct separation with high performance liquid chromatography, *Bio/Technology* **2**, 957–964.

Eisenbarth, G.S. (1985). Monoclonal antibodies, in *Comprehensive Biotechnology*, Vol. 4. Ed. Moo-Young, M., pp. 31–40. Oxford, Pergamon.

Elander, R.P. (1985). Present and future roles for biotechnology in the fermentation industry, in *Developments in Industrial Microbiology*, Vol. 26. Ed. Underkofler, L.A., pp. 1–21. Arlington, Society for Industrial Microbiology.

Enders, Jr, G.L. and Kim, H.S. (1985). AgriCultures: beneficial applications for crops and animals, in *Developments in Industrial Microbiology*, Vol. 26. Ed. Underkofler, L.A., pp. 347–376. Arlington, Society for Industrial Microbiology.

Enei, H., Shibai, H. and Hirose, Y. (1982). Amino acids and nucleic acid-related substances, in *Annual Reports on Fermentation Processes*, Vol. 5. Ed. Tsao, G.T., pp. 79–100. Orlando, Academic Press.

Enei, H., Shibai, H. and Hirose, Y. (1985). 5′-Guanosine monophosphate, in *Comprehensive Biotechnology*, Vol. 3. Ed. Moo-Young, M., pp. 653–658. Oxford, Pergamon.

Eveleigh, D.E. (1981). The microbial production of industrial chemicals, *Scientific American* **245**, 155–178.

Fassatiova, O. (1968). Moulds and filamentous fungi in technical biology, in *Progress in Industrial Microbiology*, Vol. 22. Amsterdam, Elsevier.

Fayerman, J.T. (1986). New developments in gene-cloning in antibiotic producing microorganisms, *Bio/Technology* **4**, 786–789.

Fish, N.M. and Lilly, M.D. (1984). The interactions between fermentation and protein recovery, *Bio/Technology* **2**, 623–627.

Flickinger, M. (1985). Anticancer agents, in *Comprehensive Biotechnology*, Vol. 3. Ed. Moo-Young, M., pp. 231–273. Oxford, Pergamon.

Flynn, D.S. (1983). Instrumentation and control of fermenters, in *The Filamentous Fungi*, Vol. IV, *Fungal Technology*. Eds. Smith, J.E., Berry, D.R. and Kristiansen, B., pp. 77–100. London, Arnold.

Fogarty, W.M. (1980). *Microbial Enzymes and Biotechnology.* London, Applied Science Publishers.

Fowler, M.W. and Stepan-Sarkissan, G. (1983). Chemicals from plant cell fermentation, in *Advances in Biotechnological Processes*, Vol. 2. Eds. Mizrahi, A. and Van Wezel, A.L., pp. 135–158. New York, Alan R. Liss.

Fujita, Y. and Hara, Y. (1985). The effective production of shikonin by cultures with an increased cell population, *Agricultural and Biological Chemistry* **49**(7), 2071–2075.

Furuya, A., Abe, S. and Kinoshita, S. (1968). Production of nucleic acid related substances by fermentative processes. XIX. Accumulation of 5'-inosinic acid by a mutant of *Brevibacterium ammoniagenes*, *Applied Microbiology* **16**, 981–987.

Gaden, E.L. (1981). Production methods in industrial microbiology, *Scientific American* **245**, 181–197.

Germanier, R. (1984). *Bacterial Vaccines.* Orlando, Academic.

Glick, B.R. and Whitney, G.K. (1987). Factors affecting the expression of foreign proteins in *Escherichia coli*, *Journal of Industrial Microbiology* **1**, 277–282.

Godfrey, T. and Reichelt, J. (1983). *Industrial Enzymology.* New York, Nature Press.

Greenshields, R.N. (1978). Acetic acid: vinegar, in *Economic Microbiology*, Vol. 2. Ed. Rose, A.H., pp. 121–186. London, Academic Press.

Grein, A. (1987). Antitumour anthracyclines produced by *Streptomyces peucetius*, in *Advances in Applied Microbiology*, Vol. 32. Ed. Laskin, A.I., pp. 203–214. Orlando, Academic Press.

Guthrie, R.K. and Davis, E.M. (1985). Biodegradation in effluents, in *Advances in Biotechnological Processes*, Vol. 5. Eds. Mizrahi, A. and Van Wezel, A.L., pp. 149–192. New York, Alan R. Liss.

Hamer, G. (1985). Chemical engineering and biotechnology, in *Biotechnology, Principles and Applications.* Eds. Higgins, I.J., Best, D.J. and Jones, J., pp. 346–414. Oxford, Blackwell Scientific.

Hamman, J.P. and Calton, G.J. (1985). Immunosorbent chromatography for recovery of protein products, in *Purification of Fermentation Products.* Eds. Le Roith, D., Shiloach, J. and Leahy, T.J., pp. 105–122. ACS Symposium Series 271. Washington, American Chemical Society.

Hara, Y. and Suga, C. (1986). Method for producing secondary metabolites of plants, EP0071999B1 (European patent).

Heckendorf, A.H., Ashare, E. and Rausch, C. (1985). Process scale chromatography, in *Purification of Fermentation Products.* Eds. Le Roith, D., Shiloach, J. and Leahy, T.J., pp. 91–103. ACS Symposium Series 271. Washington, American Chemical Society.

Helbert, J.R. (1982). Beer, in *Prescott and Dunn's Industrial Microbiology*, 4th Edition. Ed. Reed, G., pp. 403–467. Westport, AVI Publishing Co.

Hemming, M.L., Ousby, J.C., Plowright, D.R. and Walker, J. (1977). 'Deep shaft'—latest position, *Water Pollution Control* **76**, 441–451.

Hirose, Y., Enei, H. and Shibai, H. (1985). L-Glutamic acid fermentation, in *Comprehensive Biotechnology*, Vol. 3. Ed. Moo-Young, M., pp. 593–600. Oxford, Pergamon.

Hopwood, D.A. (1981). The genetic programming of industrial microorganisms. *Scientific American* **245**, 91–102.

Hough, J.S. (1985). *The Biotechnology of Malting and Brewing.* Cambridge, Cambridge University Press.

Hsiung, H.M., Mayne, N.G. and Becker, G.W. (1986). High-level expression, efficient screening and folding of human growth hormone in *Escherichia coli*, *Bio/Technology* **4**, 997–999.

Huggins, A.R. (1984). Progress in dairy starter cultures, *Food Technology* **38**(6), 41–50.

Hutter, R. (1982). Design of culture media capable of provoking wide gene expression, in *Bioactive Microbial Products: Search and Discovery*. Eds. Bu'Lock, J.D., Nisbet, L.J. and Winstanley, D.J., pp. 37–50. London, Academic Press.

Ignoffo, C.M. and Anderson, R.F. (1979). Bioinsecticides, in *Microbial Technology*, Vol. 1. Eds. Peppler, H.J. and Perlman, D., pp. 1–28. New York, Academic Press.

Irving, D.M. and Hill, A.R. (1985). Cheese technology, in *Comprehensive Biotechnology*, Vol. 3. Ed. Moo-Young, M., pp. 523–565. Oxford, Pergamon.

Jayme, D.W. and Blackman, K.E. (1985). Culture media for propagation of mammalian cells, viruses and other biologicals, in *Advances in Biotechnological Processes*, Vol. 5. Eds. Mizrahi, A. and Van Wezel, A.L., pp. 1–30. New York, Alan R. Liss.

Jegede, V.A., Kowal, K.J., Lin, W. and Ritchey, M.B. (1978). Vaccine technology, in *Kirk-Othmer Encyclopedia of Chemical Technology*, Vol. 23. Ed. Grayson, M. pp. 629–643. New York, Wiley.

Johnson, I.S. (1983). Human insulin from recombinant DNA technology, *Science* **219**, 632–637.

Katinger, H.W.D. and Blien, R. (1983). Production of enzymes and hormones by mammalian cell culture, in *Advances in Biotechnological Processes*, Vol. 2. Eds. Mizrahi, A. and Van Wezel, A.L., pp. 61–95. New York, Alan R. Liss.

Kato, N., Tani, Y. and Yamada, H. (1983). Microbial utilization of methanol: production of useful metabolites, in *Advances in Biotechnological Processes*, Vol. 1. Eds. Mizrahi, A. and Van Wezel, A.L., pp. 171–203. New York, Alan R. Liss.

Kennedy, J.F. and Bradshaw, I.J. (1984). Production, properties and applications of xanthan, in *Progress in Industrial Microbiology*, Vol. 19. Ed. Bushell, M.E., pp. 319–372. Amsterdam, Elsevier.

Khosrovi, B. and Gray, P.P. (1985). Products from recombinant DNA, in *Comprehensive Biotechnology*, Vol. 3. Ed. Moo-Young, M., pp. 319–330. Oxford, Pergamon.

Kisser, M., Kubicek, C.P. and Röhr, M. (1980). Influence of manganese on morphology and cell wall composition of *Aspergillus niger* during citric acid fermentation, *Archives of Microbiology* **128**, 26–33.

Klausner, A. (1986). 'Single chain' antibodies become reality, *Bio/Technology* **4**, 1041–1043.

Klein, F., Ricketts, R., Pickle, D. and Flickinger, M.C. (1983). Interleukin-2 and Interleukin-3: suspension cultures of constitutive producer cell lines, in *Advances in Biotechnological Processes*, Vol. 2. Eds. Mizrahi, A. and Van Wezel, A.L., pp. 111–134. New York, Alan R. Liss.

Klibanov, A.M. (1986). Enzymes that work in organic solvents. *Chemtech*, **16**, 354–359.

Kristiansen, B. and Chamberlain, H.E. (1983). Fermenter design, in *The Filamentous Fungi*, Vol. 4. Eds. Smith, J.E., Berry, D.R. and Kristiansen, B., pp. 1–19. London, Edward Arnold.

Kubicek, C.P. and Röhr, M. (1986). Citric acid fermentation, in *Critical Reviews of Biotechnology*, Vol. 3. Eds. Stewart, G.G. and Russell, I., pp. 331–373. Boca Raton, CRC.

Kula, M.-R. (1985). Liquid–liquid extraction of biopolymers, in *Comprehensive Biotechnology*, Vol. 2. Ed. Moo-Young, M., pp. 451–471. Oxford, Pergamon.

Laine, B.M. (1974). What proteins cost from oil, *Hydrocarbon Processing* **53**(11), pp. 139–142.

Lambert, P.W. (1983). Industrial enzyme production and recovery from filamentous fungi, in *The Filamentous Fungi*, Vol. IV, *Fungal Technology*. Eds. Smith, J.E., Berry, D.R. and Kristiansen, B., pp. 210–237. London, Edward Arnold.

Law, B.A. (1982). Cheeses, in *Fermented Foods*, *Economic Microbiology*, Vol. 7, Ed. Rose, A.H., pp. 147–198. London, Academic Press.

Law, B.A. (1984). Microorganisms and their enzymes in the maturation of cheeses, in

*Progress in Industrial Microbiology*, Vol. 19. Ed. Bushell, M.E., pp. 245–284. Amsterdam, Elsevier.

Le, M.S. and Howell, J.A. (1985). Ultrafiltration, in *Comprehensive Biotechnology*, Vol. 2. Ed. Moo-Young, M., pp. 383–409. Oxford, Pergamon.

Lechevalier, H.A. and Solotorovsky, M. (1974). *Three Centuries of Microbiology*. New York, Dover.

Lehninger, A.L. (1982). *Principles of Biochemistry*. New York, Worth Publishers.

Litchfield, J.H. (1983). Single-cell proteins, *Science* **219**, 740–746.

Lockwood, L.B. (1979). Production of organic acids by fermentation, in *Microbial Technology*, Vol. 1. Eds. Peppler, H.J. and Perlman, D., pp. 355–387. New York, Academic Press.

Lonsane, B.K., Ghildyal, N.P., Budiatman, S. and Ramakrishna, S.V. (1985). Engineering aspects of solid-state fermentation, *Enzyme and Microbial Technology* **7**, 258–265.

Lyons, T.P. and Rose, A.H. (1977). Whisky, in *Alcoholic Beverages, Economic Microbiology*, Vol. 1. Ed. Rose, A.H., pp. 635–692.

Macleod, A.M. (1977). Beer, in *Alcoholic Beverages, Economic Microbiology*, Vol. 1. Ed. Rose, A.H., pp. 43–137. London, Academic Press.

Magee, R.J. and Kosaric, N. (1987). The Microbial production of 2,3-butanediol, in *Advances in Applied Microbiology*, Vol. 32. Ed. Laskin, A.I., pp. 89–161. Orlando, Academic Press.

Margaritis, A. and Pace, G.W. (1985). Microbial polysaccharides, in *Comprehensive Biotechnology*, Vol. 3. Ed. Moo-Young, M., pp. 1005–1043. Oxford, Pergamon.

Maxon, W.D. (1985). Steroid bioconversions: one industrial perspective, in *Annual Reports on Fermentation Processes*, Vol. 8. Ed. Tsao, G.T., pp. 171–186. Orlando, Academic Press.

McGregor, W.C. (1983). Large-scale isolation and purification of recombinant proteins from recombinant *E. coli*. *Annals of the New York Academy of Science* **413**, 231–236.

McNeil, B. and Kristiansen, B. (1986). The acetone butanol fermentation, in *Advances in Applied Microbiology*, Vol. 32. Ed. Laskin, A.I., pp. 61–92. Orlando, Academic Press.

Miller, M.W. (1982). Yeasts, in *Prescott and Dunn's Industrial Microbiology*, 4th Edition. Ed. Reed, G., pp. 15–43. Westport, AVI Publishing Co.

Miller, T.L. (1985). Steroid fermentations, in *Comprehensive Biotechnology*, Vol. 3. Ed. Moo-Young, M., pp. 297–318. Oxford, Pergamon.

Miller, T.L. and Churchill, B.W. (1986). Substrates for large-scale fermentations, in *Manual of Industrial Microbiology and Biotechnology*. Eds. Demain, A.L. and Solomon, N.A., pp. 122–136. Washington, American Society for Microbiology.

Millis, N.F. (1985). The organisms of biotechnology, in *Comprehensive Biotechnology*, Vol. 3. Ed. Moo-Young, M., pp. 7–19. Oxford, Pergamon.

Misawa, M. (1985). Production of useful plant metabolites, in *Advances in Biochemical Engineering/Biotechnology*, Vol. 31. Ed. Fiechter, A., pp. 59–88. Berlin, Springer–Verlag.

Moo-Young, M., Moreira, A.R. and Tengerdy, R.P. (1983). Principles of solid-substrate fermentation, in *The Filamentous Fungi*, Vol. IV, *Fungal Technology*. Eds. Smith, J.E., Berry, D.R. and Kristiansen, B., pp. 117–144. London, Edward Arnold.

Nakayama, K. (1985). Lysine, in *Comprehensive Biotechnology*, Vol. 3. Ed. Moo-Young, M., pp. 607–620. Oxford, Pergamon.

Ng, T.K., Busche, R.M., McDonald, C.C. and Hardy, R.W.F. (1983). Production of feedstock chemicals, *Science* **219**, 733–740.

Nielsen, M.D., Henning, M.D. and Duncan, J.R. Monoclonal antibodies in veterinary medicine, in *Biotechnology and Genetic Engineering Reviews*, Vol. 1. Ed. Russel, G.E., pp. 331–353. Newcastle, Intercept.

Nilsson, K., Buzsaky, F. and Mosbach, K. (1986). Growth of anchorage-dependent cells on macroporous microcarriers, *Bio/Technology* **4**, 989–990.

Ogata, K., Kinoshita, S., Tsunoda, T. and Aida, K. (1976). *Microbial Production of Nucleic Acid-Related Substances*. Tokyo, Kodansha.

Oki, T. (1984). Recent developments in the process improvement of production of antitumour anthracycline antibiotics, in *Advances in Biotechnological Processes*, Vol. 3. Eds. Mizrahi, A. and Van Wezel, A.L., pp. 163–196. New York, Alan R. Liss.

Olsen, S. (1986). *Biotechnology, An Industry Comes of Age*. Washington, National Academy.

Onions, A.H.S., Allsopp, D. and Eggins, H.O.W. (1981). *Smith's Introduction to Industrial Mycology*, 7th Edition. London, Edward Arnold.

Onken, U. and Weiland, P. (1983). Airlift fermenters: construction, behaviour and uses, in *Advances in Biotechnological Processes*, Vol. 1. Eds. Mizrahi, A. and Van Wezel, A.L., pp. 67–95. New York, Alan R. Liss.

Oura, E., Soumalainen, H. and Viskari, R. (1982). Breadmaking, in *Fermented Foods, Economic Microbiology*, Vol. 7. Ed. Rose, A.H., pp. 87–146. London, Academic Press.

Pandrey, R.C., Kalita, C.C., Gustafson, M.E., Kline, M.C., Leidhecker, M.E. and Ross, J.T. (1985). Process developments in the isolation of largomycin F-II, a chromoprotein antitumour antibiotic, in *Purification of Fermentation Products*. Eds. Le Roith, D., Shiloach, J. and Leahy, T.J., ACS Symposium Series 271, pp. 133–153. Washington, American Chemical Society.

Phaff, H.J. (1981). Industrial microorganisms, *Scientific American* **245**, pp. 77–89.

Pirt, S.J. (1982). Microbial photosynthesis in the harnessing of solar energy, *Journal of Chemical Technology and Biotechnology* **32**, 198–202.

Porubscan, R.S. and Sellars, R.L. (1979). Lactic starter culture concentrates, in *Microbial Technology*, Vol. 1. Eds. Peppler, H.J. and Perlman, D., pp. 59–91. New York, Academic Press.

Posillico, E.G. (1986). Microencapsulation technology for large-scale antibody production, *Bio/Technology* **4**, 114–117.

Priest, F.G. (1984). *Extracellular Enzymes*. Wokingham, UK, Van Nostrand Reinhold.

Purchas, D.B. (1971). *Industrial Filtration of Liquids*. Glasgow, Leonard Hill.

Queener, S.W. and Swartz, R.W. (1979). Penicillins: biosynthetic and semisynthetic, in *Economic Microbiology*, Vol. 3. Ed. Rose, A.H., pp. 35–123. London, Academic Press.

Ratafia, M. (1987). Mammalian cell culture: worldwide activities and markets, *Bio/Technology* **5**, 692–694.

Reed, G. (1982). Microbial biomass, single cell protein, and other microbial products, in *Prescott and Dunn's Industrial Microbiology*, 4th Edition. Ed. Reed, G., pp. 541–592. Westport, AVI Publishing Co.

Reuveny, S. (1983). Microcarriers for culturing mammalian cells and their applications, in *Advances in Biotechnological Processes*, Vol. 1. Eds. Mizrahi, A. and Van Wezel, A.L., pp. 1–32. New York, Alan R. Liss.

Ricketts, R.T. and Lebherz III, W.B., Klein, F., Gustafson, M.E. and Flickinger, M.C. (1985). Application, sterilization and decontamination of ultrafiltration systems for large-scale production of biologicals, in *Purification of Fermentation Products*. Eds. Le Roith, D., Shiloach, J. and Leahy, T.J., pp. 51–69. ACS Symposium Series 271. Washington, American Chemical Society.

Robbers, J.E. (1984). The fermentative production of ergot alkaloids, in *Advances in Biotechnological Processes*, Vol. 3. Eds. Mizrahi, A. and Van Wezel, A.L., pp. 197–239. New York, Alan R. Liss.

Rolz, C., de Cabrera, S., Calzada, F., Garcia, R., de Leon, R., del Carmen de Arriola, M., de Micheo, F. and Morales, E. (1983). Concepts on the biotransformation of carbohydrates into fuel ethanol, in *Advances in Biotechnological Processes*, Vol. 1. Eds. Mizrahi, A. and Van Wezel, A.L., pp. 97–142. New York, Alan R. Liss.

Rose, A.H. (1980). *Microbial Enzymes and Bioconversions, Economic Microbiology*, Vol. 5. London, Academic Press.

Rose, A.H. (1981). The microbiology of food and drink, *Scientific American* **245**, 127–138.

Samuelov, N.S. (1983). Single-cell protein production: review of alternatives, in *Advances in Biotechnological Processes*, Vol. 1. Eds. Mizrahi, A. and Van Wezel, A.L., pp. 293–336. New York, Alan R. Liss.

Shoham, J. (1983). Production of human immune interferon, in *Advances in Biotechnological Processes*, Vol. 2. Eds. Mizrahi, A. and Van Wezel, A.L., pp. 209–269. New York, Alan R. Liss.

Sinden, K.W. (1987). The production of lipids by fermentation within the EEC, *Enzyme and Microbial Technology* **9**, 124–125.

Sitrin, R.D., Chan, G., De Phillips, P., Dingerdissen, J., Valenta, J. and Snader, K. (1985). Preparative reversed phase high performance liquid chromatography, in *Purification of Fermentation Products*. Eds. Le Roith, D., Shiloach, J. and Leahy, T.J., pp. 71–89. ACS Symposium Series 271. Washington, American Chemical Society.

Smith, J.E. (1985). *Biotechnology Principles*. Wokingham, UK, Van Nostrand Reinhold.

Solomons, G.L. (1983). Single Cell Protein, in *Critical Reviews of Biotechnology*, Vol. I. Eds. Stewart, G.G. and Russell, I., pp. 21-58. Boca Raton, CRC.

Solomons, G.L. (1985). Production of biomass by filamentous fungi, in *Comprehensive Biotechnology*, Vol. 3. Ed. Moo-Young, M., pp. 483–505. Oxford, Pergamon.

Spier, R.E. (1983). Production of veterinary vaccines, in *Advances in Biotechnological Processes*, Vol. 2. Eds. Mizrahi, A. and Van Wezel, A.L., pp. 33–59. New York, Alan R. Liss.

Stanbury, P.F. and Whitaker, A. (1984). *Principles of Fermentation Technology*. Oxford, Pergamon.

Steinkraus, K.H. (1983). Industrial applications of oriental fungal fermentations, in *The Filamentous Fungi*, Vol. IV, *Fungal Technology*, Eds. Smith, J.E., Berry, D.R. and Kristiansen, B., pp. 171–189. London, Edward Arnold.

Stewart, G.G., Panchal, C., Russell, I. and Sills, A.M. (1984). Biology of ethanol-producing organisms, in *Critical Reviews of Biotechnology*, Vol. 1. Eds. Stewart, G.G. and Russell, I., pp. 161–188. Boca Raton, CRC.

Stewart, G.G. and Russell, I. (1985). Modern brewing technology, in *Comprehensive Biotechnology*, Vol. 3. Ed. Moo-Young, M., pp. 335–382. Oxford, Pergamon.

Street, G. (1983). Large scale industrial enzyme production, in *Critical Review of Biotechnology*, Vol. 1. Eds. Stewart, G.G. and Russell, I., pp. 59–89. Boca Raton, CRC.

Strom, P.F. and Chung, J.-C. (1985). The rotating biological contactor for wastewater treatment, in *Advances in Biotechnological Processes*, Vol. 5. Eds. Mizrahi, A. and Van Wezel, A.L., pp. 193–225. New York, Alan R. Liss.

Stroshane, R.M. (1984). Production of daunorubicin, in *Advances in Biotechnological Processes*, Vol. 3. Eds. Mizrahi, A. and Van Wezel, A.L., pp. 141–161, New York, Alan R. Liss.

Stutzenberger, F. (1985). Regulation of cellulolytic activity, in *Annual Reports on Fermentation Processes*, Vol. 8. Ed. Tsao, G.T., pp. 111–154. Orlando, Academic Press.

Swartz, R.W. (1985). Penicillins, in *Comprehensive Biotechnology*, Vol. 3. Ed. Moo-Young, M., pp. 7–47. Oxford, Pergamon.

Szmant, H.H. (1986). *Industrial Utilization of Renewable Resources*. Lancaster, Technomic.

Tautorus, T.E. (1985). Mushroom fermentation, in *Advances in Biotechnological Processes*, Vol. 5. Eds. Mizrahi, A. and Van Wezel, A.L., pp. 227–273. New York, Alan R. Liss.

Thielsch, H. (1967). Manufacture, fabrication and joining of commercial piping, in *Piping Handbook*. Ed. King, R.C., pp. 7.1–7.300. New York, McGraw-Hill.

Trevan, M.D., Boffey, S., Goulding, K.H. and Stanbury, P. (1987). *Biotechnology, The Biological Principles*. Milton Keynes, Open University Press.

Tutunjian, R.S. (1985). Ultrafiltration processes, in *Comprehensive Biotechnology*, Vol. 2. Ed. Moo-Young, M., pp. 411–437. Oxford, Pergamon.

Tzeng, C.H. (1985). Applications for starter cultures in the dairy industry, in *Developments in Industrial Microbiology*, Vol. 26. Ed. Underkofler, L.A., pp. 323–338. Arlington, Society for Industrial Microbiology.

Van Brunt, J. (1986a). Fungi: the perfect hosts? *Bio/Technology* 4, 1057–1062.

Van Brunt, J. (1986b). Immobilized mammalian cells: the gentle way to productivity, *Bio/Technology* 4, 505–510.

Van Hemert, P. (1974). Vaccine production as a unit process, in *Progress in Industrial Microbiology*, Vol. 13. Ed. Hockenhull, D.J.D., pp. 151–271. Edinburgh, Churchill Livingstone.

Van Uden, N. (1985). Ethanol toxicity and ethanol tolerance of yeasts, in *Annual Reports on Fermentation Processes*, Vol. 8. Ed. Tsao, G.T., pp. 11–58. Orlando, Academic Press.

Ward, O.P. (1985a). Hydrolytic enzymes, in *Comprehensive Biotechnology*, Vol. 3. Ed. Moo-Young, M., pp. 819–836.

Ward, O.P. (1985b). Proteolytic enzymes, in *Comprehensive Biotechnology*, Vol. 3. Ed. Moo-Young, M., pp. 789–818. Oxford, Pergamon.

Wasserman, B.P. (1984). Thermostable enzyme production, *Food Technology* 38 (2), 78–89, 98.

Watson, J.D., Hopkins, N.H., Roberts, J.W., Steitz, J.A. and Weiner, A.M. (1987). *Molecular Biology of the Gene*, 4th Edition. California, Benjamin/Cummings.

Wheatley, A.D. (1984). Biotechnology of effluent treatment, in *Biotechnology and Genetic Engineering Reviews*, Vol. 1. Ed. Russell, G.E., pp. 261–310. Newcastle, Intercept.

White, R.J., Klein, F., Chan, J.A. and Stroshane, R.M. (1980). Large-scale production of human interferons, in *Annual Reports on Fermentation Processes*, Vol. 4. Ed. Tsao, G.T., pp. 109–234.

White, T.J., Meade, J.H., Shoemaker, S.P., Koths, K.E. and Innis, M.A. (1984). Enzyme cloning for the food fermentation industry, *Food Technology* 38 (2), 90–95.

Wiegel, J. and Ljungdahl, L. (1986). The importance of thermophilic bacteria in biotechnology, in *Critical Reviews of Biotechnology*, Vol. 3. Eds. Stewart, G.G. and Russell, I., pp. 39–108. Boca Raton, CRC.

Wilson, T. (1984). Bioreactor, synthesizer, biosensor markets to increase by 16 percent annually, *Bio/Technology* 2, 869–873.

Wiseman, A. (1983). *Principles of Biotechnology*. London, Surrey University Press.

Wodzinski, R.J., Gennaro, R.N. and Scholla, M.H. (1987). Economics of the bioconversion of biomass to methane and other vendable products, in *Advances in Applied Microbiology*, Vol. 32. Ed. Laskin, A.I., pp. 37–88. Orlando, Academic Press.

Wood, B.J.B. (1982). Soy Sauce and Miso, in *Fermented Foods, Economic Microbiology*, Vol. 7. Ed. Rose, A.H., pp. 39–86. London, Academic Press.

Wood, B.J.B. (1984). Progress in soy sauce and related fermentations, in *Progress in Industrial Microbiology*, Vol. 19. Ed. Bushell, M.E., pp. 373–410. Amsterdam, Elsevier.

Workman, W.E., McLinden, J.H. and Dean, D.H. (1986). Genetic engineering applications to biotechnology in the genus *Bacillus*, in *Critical Reviews of Biotechnology*, Vol. 3. Eds. Stewart, G.G. and Russell, I., pp. 199–234. Boca Raton, CRC.

Zahka, J. and Leahy, T.J. (1985). Practical aspects of tangential flow filtration in cell separations, in *Purification of Fermentation Products*, Eds. Le Roith, D., Shiloach, J. and Leahy, T.J., pp. 51–69. ACS Symposium Series 271. Washington, American Chemical Society.

# Index